An Introduction to Waldorf Education and Other Essays

An Introduction to Waldorf Education and Other Essays

by Rudolf Steiner

Wilder Publications, LLC.
PO Box 3005
Radford VA 24143-3005

ISBN 10: 1-5154-3626-8
ISBN 13: 978-1-5154-3626-3

Table of Contents

An Introduction to Waldorf Education

The aims Emil Molt is trying to realize through the Waldorf School are connected with quite definite views on the social tasks of the present day and the near future. The spirit in which the school should be conducted must proceed from these views. It is a school attached to an industrial undertaking. The peculiar place modern industry has taken in the evolution of social life in actual practice sets its stamp upon the modern social movement. Parents who entrust their children to this school are bound to expect that the children shall be educated and prepared for the practical work of life in a way that takes due account of this movement. This makes it necessary, in founding the school, to begin from educational principles that have their roots in the requirements of modern life. Children must be educated and instructed in such a way that their lives fulfill demands everyone can support, no matter from which of the inherited social classes one might come. What is demanded of people by the actualities of modern life must find its reflection in the organization of this school. What is to be the ruling spirit in this life must be aroused in the children by education and instruction.

It would be fatal if the educational views upon which the Waldorf School is founded were dominated by a spirit out of touch with life. Today, such a spirit may all too easily arise because people have come to feel the full part played in the recent destruction of civilization by our absorption in a materialistic mode of life and thought during the last few decades. This feeling makes them desire to introduce an idealistic way of thinking into the management of public affairs. Anyone who turns his attention to developing educational life and the system of instruction will desire to see such a way of thinking realized there especially. It is an attitude of mind that reveals much good will. It goes without saying that this good will should be fully appreciated. If used properly, it can provide valuable service when gathering manpower for a social undertaking requiring new foundations. Yet it is necessary in this case to point our how the best intentions must fail if they set to work without fully regarding those first conditions that are based on practical insight.

This, then, is one of the requirements to be considered when the founding of any institution such as the Waldorf School is intended. Idealism must work in the spirit of its curriculum and methodology; but it must be an idealism that has the power to awaken in young, growing human beings the forces and faculties they will need in later life to be equipped for work in modern society and to obtain for themselves an adequate living.

The pedagogy and instructional methodology will be able to fulfill this requirement only through a genuine knowledge of the developing human being. Insightful people are today calling for some form of education and instruction directed not merely to the cultivation of one-sided knowledge, but also to abilities; education directed not merely to the cultivation of intellectual faculties, but also to the strengthening of the will. The soundness of this idea is unquestionable; but it is impossible to develop the will (and that healthiness of feeling on which it rests) unless one develops the insights that awaken the energetic impulses of will and feeling. A mistake often made presently in this respect is not that people instill too many concepts into young minds, but that the kind of concepts they cultivate are devoid of all driving life force. Anyone who believes one can cultivate the will without cultivating the concepts that give it life is suffering from a delusion. It is the business of contemporary educators to see this point clearly; but this clear vision can only proceed from a living understanding of the whole human being.

It is now planned that the Waldorf School will be a primary school in which the educational goals and curriculum are founded upon each teacher's living insight into the nature of the whole human being, so far as this is possible under present conditions. Children will, of course, have to be advanced far enough in the different school grades to satisfy the standards imposed by the current views. Within this framework, however, the pedagogical ideals and curriculum will assume a form that arises out of this knowledge of the human being and of actual life.

The primary school is entrusted with the child at a period of its life when the soul is undergoing a very important transformation. From birth to about the sixth or seventh year, the human being naturally gives himself up to everything immediately surrounding him in the human environment, and thus, through the imitative instinct, gives form to his own nascent powers. From this period on, the child's soul becomes open to take in consciously what the educator and teacher gives, which affects the child as a result of the teacher's natural authority. The authority is taken for granted by the child from a dim feeling that in the teacher there is something that should exist in himself, too. One cannot be an educator or teacher unless one adopts out of full insight a stance toward the child that takes account in the most comprehensive sense of this metamorphosis of the urge to imitate into an ability to assimilate upon the basis of a natural relationship of authority. The

modern world view, based as it is upon natural law, does not approach these
fact of human development in full consciousness. To observe them with the
necessary attention, one must have a sense of life's subtlest manifestations in
the human being. This kind of sense must ran through the whole an of
education; it must shape the curriculum; it must live in the spirit uniting
teacher and pupil. In educating, what the teacher does can depend only
slightly on anything he gets from a general, abstract pedagogy: it must rather
be newly born every moment from a live understanding of the young human
being he or she is teaching. One may, of course, object that this Lively kind
of education and instruction breaks down in large classes. This objection is
no doubt justified in a limited sense. Taken beyond those limits, however, the
objection merely shows that the person who makes it proceeds from abstract
educational norms, for a really living an of education based on a genuine
knowledge of the human being carries with it a power that rouses the interest
of every single pupil so that there is no need for direct "individual" work in
order to keep his attention on the subject. One can put forth the essence of
one's teaching in such a form that each pupil assimilates it in his own
individual way. This requires simply that whatever the teacher does should
be sufficiently alive. If anyone has a genuine sense for human nature, the
developing human being becomes for him such an intense, Living riddle that
the very attempt to solve it awakens the pupil's living interest empathetically.
Such empathy is more valuable than individual work, which may all too easily
cripple the child's own initiative. It might indeed be asserted — again, within
limitations — that large classes led by teachers who are imbued with the life
that comes from genuine knowledge of the human being, will achieve better
results than small classes led by teachers who proceed from standard
educational theories and have no chance to put this life into their work.

Not so outwardly marked as the transformation the soul undergoes in the
sixth or seventh year, but no less important for the art of educating, is a
change that a penetrating study of the human being shows to take place
around the end of the ninth year. At this time, the sense of self assumes a
form that awakens in the child a relationship to nature and to the world
about him such that one can now talk to him more about the connections
between things and processes themselves, whereas previously he was
interested almost exclusively in things and processes only in relationship to
man. Facts of this kind in a human being's development ought to be most
carefully observed by the educator. For if one introduces into the child's
world of concepts and feelings what coincides just at that period of life with
the direction taken by his own developing powers, one then gives such added

vigor to the growth of the whole person that it remains a source of strength throughout life. If in any period of life one works against the grain of these developing powers, one weakens the individual.

Knowledge of the special needs of each life period provides a basis for drawing up a suitable curriculum. This knowledge also can be a basis for dealing with instructional subjects in successive periods. By the end of the ninth year, one must have brought the child to a certain level in all that has come into human life through the growth of civilization. Thus while the first school years are properly spent on teaching the child to write and read, the teaching must be done in a manner that permits the essential character of this phase of development to be served. If one teaches things in a way that makes a one-sided claim on the child's intellect and the merely abstract acquisition of skills, then the development of the native will and sensibilities is checked; while if the child learns in a manner that calls upon its whole being, he or she develops all around. Drawing in a childish fashion, or even a primitive kind of painting, brings out the whole human being's interest in what he is doing. Therefore one should let writing grow out of drawing. One can begin with figures in which the pupil's own childish artistic sense comes into play; from these evolve the letters of the alphabet. Beginning with an activity that, being artistic, draws out the whole human being, one should develop writing, which tends toward the intellectual. And one must let reading, which concentrates the attention strongly within the realm of the intellect, arise out of writing.

When people recognize how much is to be gained for the intellect from this early artistic education of the child, they will be willing to allow art its proper place in the primary school education. The arts of music, painting and sculpting will be given a proper place in the scheme of instruction. This artistic element and physical exercise will be brought into a suitable combination. Gymnastics and action games will be developed as expressions of sentiments called forth by something in the nature of music or recitation. Eurythmic movement — movement with a meaning — will replace those motions based merely on the anatomy and physiology of the physical body. People will discover how great a power resides in an artistic manner of instruction for the development of will and feeling. However, to teach or instruct in this way and obtain valuable results can be done only by teachers who have an insight into the human being sufficiently keen to perceive clearly the connection between the methods they are employing and the developmental forces that manifest themselves in any particular period of life. The real teacher, the real educator, is not one who has studied educational theory as a science of the management of children, but one in whom the pedagogue has been awakened by awareness of human nature.

Of prime importance for the cultivation of the child's feeling-life is that the child develops its relationship to the world in a way such as that which develops when we incline toward fantasy. If the educator is not himself a fantast, then the child is not in danger of becoming one when the teacher conjures forth the realms of plants and animals, of the sky and the stars in the soul of the child in fairy-tale fashion.

Visual aids are undoubtedly justified within certain limits; but when a materialistic conviction leads people to try to extend this form of teaching to every conceivable thing, they forget there are other powers in the human being which must be developed, and which cannot be addressed through the medium of visual observation. For instance, there is the acquisition of certain things purely through memory that is connected to the developmental forces at work between the sixth or seventh and the fourteenth year of life. It is this property of human nature upon which the teaching of arithmetic should be based. Indeed, arithmetic can be used to cultivate the faculty of memory. If one disregards this fact, one may perhaps be tempted (especially when teaching arithmetic) to commit the educational blunder of teaching with visual aids what should be taught as a memory exercise.

One may fall into the same mistake by trying all too anxiously to make the child understand everything one tells him. The will that prompts one to do so is undoubtedly good, but does not duly estimate what it means when, Later in life, we revive within our soul something that we acquired simply through memory when younger and now find, in our mature years, that we have come to understand it on our own. Here, no doubt, any fear of the pupil's not taking an active interest in a lesson learned by memory alone will have to be relieved by the teacher's lively way of giving it. If the teacher engages his or her whole being in teaching, then he may safely bring the child things for which the full understanding will come when joyfully remembered in later life. There is something that constantly refreshes and strengthens the inner substance of life in this recollection. If the teacher assists such a strengthening, he will give the child a priceless treasure to take along on life's road. In this way, too, the teacher will avoid the visual aid's degenerating into the banality that occurs when a lesson is overly adapted to the child's understanding. Banalities may be calculated to arouse the child's own activity, but such fruits lose their flavor with the end of childhood. The flame enkindled in the child from the living fire of the teacher in matters that still lie, in a way, beyond his "understanding," remains an active, awakening force throughout the child's life.

If, at the end of the ninth year, one begins to choose descriptions of natural history from the plant and animal world, treating them in a way that the natural forms and processes lead to an understanding of the human form and the phenomena of human life, then one can help release the forces that at this age are struggling to be born out of the depths of human nature. It is consistent with the character of the child's sense of self at this age to see the qualities that nature divides among manifold species of the plant and animal kingdoms as united into one harmonious whole at the summit of the natural world in the human being.

Around the twelfth year, another turning point in the child's development occurs. He becomes ripe for the development of the faculties that lead him in a wholesome way to the comprehension of things that must be considered without any reference to the human being: the mineral kingdom, the physical world, meteorological phenomena, and so on.

The best way to lead then from such exercises, which are based entirely on the natural human instinct of activity without reference to practical ends, to others that shall be a sort of education for actual work, will follow from knowledge of the character of the successive periods of life. What has been said here with reference to particular parts of the curriculum may be extended to everything that should be taught to the pupil up to his fifteenth year.

There need be no fear of the elementary schools releasing pupils in a state of soul and body unfit for practical life if their principles of education and instructions are allowed to proceed, as described, from the inner development of the human being. For human life itself is shaped by this inner development; and one can enter upon life in no better way than when, through the development of our own inner capacities, we can join with what others before us, from similar inner human capacities, have embodied in the evolution of the civilized world. It is true that to bring the two into harmony — the development of the pupil and the development of the civilized world — will require a body of teachers who do not shut themselves up in an educational routine with strictly professional interests, but rather take an active interest in the whole range of life. Such a body of teachers will discover how to awaken in the upcoming generation a sense of the inner, spiritual substance of life and also an understanding of life's practicalities. If instruction is carried on this way, the young human being at the age of fourteen or fifteen will not lack comprehension of important things in agriculture and industry, commerce and travel, which help to make up the collective life of mankind. He will have acquired a knowledge of things and a practical skill that will enable him to feel at home in the life which receives him into its stream.

If the Waldorf School is to achieve the aims its founder has in view, it must be built on educational principles and methods of the kind here described. It will then be able to give the kind of education that allows the pupil's body to develop healthily and according to its needs, because the soul (of which this body is the expression) is allowed to grow in a way consistent with the forces of its development. Before its opening, some preparatory work was attempted with the teachers so that the school might be able to work toward the proposed aim. Those concerned with the management of the school believe that in pursuing this aim they bring something into educational life in accordance with modern social thinking. They feel the responsibility inevitably connected with any such attempt; but they think that, in contemporary social demands, it is a duty to undertake this when the opportunity is afforded.

Individualism in Philosophy

If the human being were a mere creature of nature and not a creator at the same time, he would not stand questioningly before the phenomena of the world and would also not seek to fathom their essential being and laws. He would satisfy his drive to eat and to propagate in accordance with the inborn laws of his organism and otherwise allow the events of the world to take the course they happen to take. It would not occur to him at all to address a question to nature. Content and happy he would go through life like the rose of which Angelus Silesius says:

The rose has no "wherefore?"; it blooms because it blooms. It pays itself no mind, asks not if it is seen.

The rose can just be like this. What it is it is because nature has made it this way. But the human being cannot just be like this. There is a drive within him to add to the world lying before him yet another world that springs forth from him. He does not want to live with his fellowmen in the chance proximity into which nature has placed him; he seeks to regulate the way he lives with others in accordance with his reason. The form in which nature has shaped man and woman does not suffice for him; he creates the ideal figures of Greek sculpture. To the natural course of events in daily life he adds the course of events springing from his imagination as tragedy and comedy. In architecture and music, creations spring from his spirit that are hardly reminiscent at all of anything created by nature. In his sciences he draws up conceptual pictures through which the chaos of world phenomena passing daily before our senses appears to us as a harmoniously governed whole, as a structured organism. In the world of his own deeds, he creates a particular realm — that of historical happenings — which is essentially different from nature's course of events.

The human being feels that everything he creates is only a continuation of the workings of nature. He also knows that he is called upon to add something higher to what nature can do out of itself. He is conscious of the fact that he gives birth out of himself to another, higher nature in addition to outer nature.

Thus the human being stands between two worlds; between the world that presses in upon him from outside and the world that he brings forth out of himself. His effort is to bring these two worlds into harmony. For, his whole being aims at harmony. He would like to live like the rose that does not ask about the whys and wherefores but rather blooms because it blooms. Schiller demands this of the human being in the words:

Are you seeking the highest, the greatest? The plant can teach it to you. What it is will-lessly, you must be will-fully — that's it!

The plant can just be what it is. For no new realm springs forth from it, and therefore the fearful longing can also not arise in it: How am I to bring the two realms into harmony with each other?

The goal for which man has striven throughout all the ages of history is to bring what lies within him into harmony with what nature creates out of itself. The fact that he himself is fruitful becomes the starting point for his coming to terms with nature; this coming to terms forms the content of his spiritual striving.

There are two ways of coming to terms with nature. The human being either allows outer nature to become master over his inner nature, or he subjects this outer nature to himself. In the first case, he seeks to submit his own willing and existence to the outer course of events. In the second case, he draws the goal and direction of his willing and existence from himself and seeks to deal in some way or other with the events of nature that still go their own way.

Let us speak about the first case first. It is in accordance with his essential being for man, above and beyond the realm of nature, to create yet another realm that in his sense is a higher one. He can do no other. How he relates to the outer world will depend upon the feelings and emotions he has with respect to this his own realm. Now he can have the same feelings with respect to his own realm as he has with respect to the facts of nature. He then allows the creations of his spirit to approach him in the same way he allows an event of the outer world, wind and weather, for example, to approach him. He perceives no difference in kind between what occurs in the outer world and what occurs within his soul. He therefore believes that they are only one realm, i.e., governed by one kind of law. But he does feel that the creations of his spirit are of a higher sort. He therefore places them above the creations of mere nature. Thus he transfers his own creations into the outer world and lets nature be governed by them. Consequently he knows only an outer world. For he transfers his own inner world outside himself. No wonder then that for him even his own self becomes a subordinate part of this outer world.

One way man comes to terms with the outer world consists, therefore, in his regarding his inner being as something outer; he sets this inner being, which he has transferred into the outer world, both over nature and over himself as ruler and lawgiver.

This characterizes the standpoint of the religious person. A divine world order is a creation of the human spirit. But the human being is not clear about the fact that the content of this world order has sprung from his own spirit. He therefore transfers it outside himself and subordinates himself to his own creation.

The acting human being is not content simply to act. The flower blooms because it blooms. It does not ask about whys and wherefores. The human being relates to what he does. He connects feelings to what he does. He is either satisfied or dissatisfied with what he does. He makes value judgments about his actions. He regards one action as pleasing to him, and another as displeasing. The moment he feels this, the harmony of the world is disturbed for him. He believes that the pleasing action must bring about different consequences than one which evokes his displeasure. Now if he is not clear about the fact that, out of himself, he has attached the value judgments to his actions, he will believe that these values are attached to his actions by some outer power. He believes that an outer power differentiates the happenings of this world into ones that are pleasing and therefore good, and ones that are displeasing and therefore bad, evil. A person who feels this way makes no distinction between the facts of nature and the actions of the human being. He judges both from the same point of view. For him the whole cosmos is one realm, and the laws governing this realm correspond entirely to those which the human spirit brings forth out of itself.

This way of coming to terms with the world reveals a basic characteristic of human nature. No matter how unclear the human being might be about his relationship to the world, he nevertheless seeks within himself the yardstick by which to measure all things. Out of a kind of unconscious feeling of sovereignty he decides on the absolute value of all happenings. No matter how one studies this, one finds that there are countless people who believe themselves governed by gods; there are none who do not independently, over the heads of the gods, judge what pleases or displeases these gods. The religious person cannot set himself up as the lord of the world; but he does indeed determine, out of his own absolute power, the likes and dislikes of the ruler of the world.

One need only look at religious natures and one will find my assertions confirmed. What proclaimer of gods has not at the same time determined quite exactly what pleases these gods and what is repugnant to them? Every religion has its wise teachings about the cosmos, and each also asserts that its wisdom stems from one or more gods.

If one wants to characterize the standpoint of the religious person one must say: He seeks to judge the world out of himself, but he does not have the courage also to ascribe to himself the responsibility for this judgment; therefore he invents beings for himself in the outer world that he can saddle with this responsibility.

Such considerations seem to me to answer the question:

What is religion? The content of religion springs from the human spirit. But the human spirit does not want to acknowledge this origin to itself. The human being submits himself to his own laws, but he regards these laws as foreign. He establishes himself as ruler over himself. Every religion establishes the human "I" as regent of the world. Religion's being consists precisely in this, that it is not conscious of this fact. It regards as revelation from outside what it actually reveals to itself.

The human being wishes to stand at the topmost place in the world. But he does not dare to pronounce himself the pinnacle of creation. Therefore he invents gods in his own image and lets the world be ruled by them. When he thinks this way, he is thinking religiously.

Philosophical thinking replaces religious thinking. Wherever and whenever this occurs, human nature reveals itself to us in a very particular way.

For the development of Western thinking, the transition from the mythological thinking of the Greeks into philosophical thinking is particularly interesting. I would now like to present three thinkers from that time of transition: Anaximander, Thales, and Parmenides. They represent three stages leading from religion to philosophy.

It is characteristic of the first stage of this path that divine beings, from whom the content taken from the human "I" supposedly stems, are no longer acknowledged. But from habit one still holds fast to the view that this content stems from the outer world. Anaximander stands at this stage. He no longer speaks of gods as his Greek ancestors did. For him the highest principle, which rules the world, is not a being pictured in man's image. It is an impersonal being, the apeiron, the indefinite. It develops out of itself everything occurring in nature, not in the way a person creates, but rather out of natural necessity. But Anaximander always conceives this natural necessity to be analogous to actions that proceed according to human principles of reason. He pictures to himself, so to speak, a moral, natural lawfulness, a highest being, that treats the world like a human, moral judge without actually being one. For Anaximander, everything in the world occurs just as necessarily as a magnet attracts iron, but does so according to moral, i.e., human laws. Only from this point of view could he say: "Whence things arise, hence must they also pass away, in accordance with justice, for they must do penance and recompense because of unrighteousness in a way corresponding to the order of time."

This is the stage at which a thinker begins to judge philosophically. He lets go of the gods. He therefore no longer ascribes to the gods what comes from man. But he actually does nothing more than transfer onto something impersonal the characteristics formerly attributed to divine, i.e., personal beings.

Thales approaches the world in an entirely free way. Even though he is a few years older than Anaximander, he is philosophically much more mature. His way of thinking is no longer religious at all.

Within Western thinking Thales is the first to come to terms with the world in the second of the two ways mentioned above. Hegel has so often emphasized that thinking is the trait which distinguishes man from the animal. Thales is the first Western personality who dared to assign to thinking its sovereign position. He no longer bothered about whether gods have arranged the world in accordance with the order of thought or whether an apeiron directs the world in accordance with thinking. He only knew that he thought, and assumed that, because he thought, he also had a right to explain the world to himself in accordance with his thinking. Do not underestimate this standpoint of Thales! It represents an immense disregard for all religious preconceptions. For it was the declaration of the absoluteness of human thinking. Religious people say: The world is arranged the way we think it to be because God exists. And since they conceive of God in the image of man, it is obvious that the order of the world corresponds to the order of the human head. All that is a matter of complete indifference to Thales. He thinks about the world. And by virtue of his thinking he ascribes to himself the power to judge the world. He already has a feeling that thinking is only a human action; and accordingly he undertakes to explain the world with the help of this purely human thinking. With Thales the activity of knowing (das Erkennen) now enters into a completely new stage of its development. It ceases to draw its justification from the fact that it only copies what the gods have already sketched out. It takes from out of itself the right to decide upon the lawfulness of the world. What matters, to begin with, is not at all whether Thales believed water or anything else to be the principle of the world; what matters is that he said to himself: What the principle is, this I will decide by my thinking. He assumed it to be obvious that thinking has the power in such things. And therein lies his greatness.

Just consider what was accomplished. No less an event than that spiritual power over world phenomena was given to man. Whoever trusts in his thinking says to himself: No matter how violently the waves of life may rage, no matter that the world seems a chaos: I am at peace, for all this mad commotion does not disquiet me, because I comprehend it.

Heraclitus did not comprehend this divine peacefulness of the thinker who understands himself. He was of the view that all things are in eternal flux. That becoming is the essential beings of things. When I step into a river, it is no longer the same one as in the moment of my deciding to enter it. But Heraclitus overlooks just one thing. Thinking preserves what the river bears along with itself and finds that in the next moment something passes before my senses that is essentially the same as what was already there before.

Like Thales, with his firm belief in the power of human thinking, Heraclitus is a typical phenomenon in the realm of those personalities who come to terms with the most significant questions of existence. He does not feel within himself the power to master by thinking the eternal flux of sense-perceptible becoming. Heraclitus looks into the world and it dissolves for him into momentary phenomena upon which one has no hold. If Heraclitus were right, then everything in the world would flutter away, and in the general chaos the human personality would also have to disintegrate. I would not be the same today as I was yesterday, and tomorrow I would be different than today. At every moment, the human being would face something totally new and would be powerless. For, it is doubtful that the experiences he has acquired up to a certain day can guide him in dealing with the totally new experiences that the next day will bring.

Parmenides therefore sets himself in absolute opposition to Heraclitus. With all the one-sidedness possible only to a keen philosophical nature, he rejected all testimony brought by sense perception. For, it is precisely this ever-changing sense world that leads one astray into the view of Heraclitus. Parmenides therefore regarded those revelations as the only source of all truth which well forth from the innermost core of the human personality: the revelations of thinking. In his view the real being of things is not what flows past the senses; it is the thoughts, the ideas, that thinking discovers within this stream and to which it holds fast!

Like so many things that arise in opposition to a particular one-sidedness, Parmenides's way of thinking also became disastrous. It ruined European thinking for centuries. It undermined man's confidence in his sense perception. Whereas an unprejudiced, naive look at the sense world draws from this world itself the thought-content that satisfies the human drive for knowledge, the philosophical movement developing in the sense of Parmenides believed it had to draw real truth only out of pure, abstract thinking.

The thoughts we gain in living intercourse with the sense world have an individual character; they have within themselves the warmth of something experienced. We unfold our own personality by extracting ideas from the world. We feel ourselves as conquerors of the sense world when we capture it in the world of thoughts. Abstract, pure thinking has something impersonal and cold about it. We always feel a compulsion when we spin forth ideas out of pure thinking. Our feeling of self cannot be heightened through such thinking. For we must simply submit to the necessities of thought.

Parmenides did not take into account that thinking is an activity of the human personality. He took it to be impersonal, as the eternal content of existence. What is thought is what exists, he once said.

In the place of the old gods he thus set a new one. Whereas the older religious way of picturing things had set the whole feeling, willing, and thinking man as God at the pinnacle of the world, Parmenides took one single human activity, one part, out of the human personality and made a divine being out of it.

In the realm of views about the moral life of man Parmenides is complemented by Socrates. His statement that virtue is teachable is the ethical consequence of Parmenides's view that thinking is equitable with being. If this is true, then human action can claim to have raised itself to something worthily existing only when human action flows from thinking, from that abstract, logical thinking to which man must simply yield himself, i.e., which he has to acquire for himself as learner.

It is clear that a common thread can be traced through the development of Greek thought. The human being seeks to transfer into the outer world what belongs to him, what springs from his own being, and in this way to subordinate himself to his own being. At first he takes the whole fullness of his nature and sets likenesses of it as gods over himself; then he takes one single human activity, thinking, and sets it over himself as a necessity to which he must yield. That is what is so remarkable in the development of man, that he unfolds his powers, that he fights for the existence and unfolding of these powers in the world, but that he is far from being able to acknowledge these powers as his own.

One of the greatest philosophers of all time has made this great, human self-deception into a bold and wonderful system. This philosopher is Plato. The ideal world, the inner representations that arise around man within his spirit while his gaze is directed at the multiplicity of outer things, this becomes for Plato a higher world of existence of which that multiplicity is only a copy. "The things of this world which our senses perceive have no true being at all: they are always becoming but never are. They have only a relative existence; they are, in their totality, only in and through their relationship to each other; one can therefore just as well call their whole existence a non-existence. They are consequently also not objects of any actual knowledge. For, only about what is, in and for itself and always in the same way, can there be such knowledge; they, on the other hand, are only the object of what we, through sensation, take them to be. As long as we are limited only to our perception of them, we are like people who sit in a dark cave so firmly bound that they cannot even turn their heads and who see nothing, except, on the wall facing them, by the light of a fire burning behind them, the shadow images of real things which are led across between them and the fire, and who in fact also see of each other, yes each of himself, only the shadows on that wall. Their wisdom, however, would be to predict the sequence of those shadows which they have learned to know from experience." The tree that I see and touch, whose flowers I smell, is therefore the shadow of the idea of the tree. And this idea is what is truly real. The idea, however, is what lights up within my spirit when I look at the tree. What I perceive with my senses is thus made into a copy of what my spirit shapes through the perception.

Everything that Plato believes to be present as the world of ideas in the beyond, outside things, is man's inner world. The content of the human spirit, torn out of man and pictured as a world unto itself, as a higher, true world lying in the beyond: that is Platonic philosophy.

I consider Ralph Waldo Emerson to be right when he says: "Among books, Plato only is entitled to Omar's fanatical compliment to the Koran, when he said, 'Burn the libraries; for their value is in this book.' These sentences contain the culture of nations; these are the cornerstone of schools; these are the fountain-head of literatures. A discipline it is in logic, arithmetic, taste, symmetry, poetry, language, rhetoric, ontology, morals, or practical wisdom. There was never such range of speculation. Out of Plato come all things that are still written and debated among men of thought." Let me express the last sentence somewhat more exactly in the following form. The way Plato felt about the relationship of the human spirit to the world, this is how the overwhelming majority of people still feel about it today. They feel that the content of the human spirit — human feeling, willing, and thinking — does stand at the top of the ladder of phenomena; but they know what to do with this spiritual content only when they conceive of it as existing outside of man as a divinity or as some other kind of higher being such as a necessary natural order, or as a moral world order — or as any of the other names that man has given to what he himself brings forth.

One can understand why the human being does this. Sense impressions press in upon him from outside. He sees colors and hears sounds. His feelings and thoughts arise in him as he sees the colors and hears the sounds. These stem from his own nature. He asks himself: How can I, out of myself, add anything to what the world gives me? It seems to him completely arbitrary to draw something out of himself to complement the outer world.

But the moment he says to himself: What I am feeling and thinking, this I do not bring to the world out of myself; another, higher being has laid this into the world, and I only draw it forth from the world — at this moment he feels relieved. One only has to tell the human being: Your opinions and thoughts do not come from yourself; a god has revealed them to you — then he is reconciled with himself. And if he has divested himself of his belief in God, he then sets in His place the natural order of things, eternal laws. The fact that he cannot find this God, these eternal laws, anywhere outside in the world, that he must rather first create them for the world if they are to be there — this he does not want to admit to himself at first. It is difficult for him to say to himself: The world outside me is not divine; by virtue of my essential being, however, I assume the right to project the divine into the outer world.

What do the laws of the pendulum that arose in Galileo's spirit as he watched the swinging church lamp matter to the lamp? But man himself cannot exist without establishing a relationship between the outer world and the world of his inner being. His spiritual life is a continuous projecting of his spirit into the sense world. Through his own work, in the course of historical life, there occurs the interpenetration of nature and spirit. The Greek thinkers wanted nothing more than to believe that man was already born into a relationship which actually can come about only through himself. They did not want it to be man who first consummates the marriage of spirit and nature; they wanted to confront this as a marriage already consummated, to regard it as an accomplished fact.

Aristotle saw what is so contradictory in transferring the ideas — arising in man's spirit from the things of the world — into some supersensible world in the beyond. But even he did not recognize that things first receive their ideal aspect when man confronts them and creatively adds this aspect to them. Rather, he assumed that this ideal element, as entelechy, is itself at work in things as their actual principle. The natural consequence of this basic view of his was that he traced the moral activity of man back to his original, moral, natural potential. The physical drives ennoble themselves in the course of human evolution and then appear as willing guided by reason. Virtue consists in this reasonable willing.

Taken at face value, this seems to indicate that Aristotle believed that moral activity, at least, has its source in man's own personality, that man himself gives himself the direction and goal of his actions out of his own being and does not allow these to be prescribed for him from outside. But even Aristotle does not dare to stay with this picture of a human being who determines his own destiny for himself. What appears in man as individual, reasonable activity is, after all, only the imprint of a general world reason existing outside of him. This world reason does realize itself within the individual person, but has its own independent, higher existence over and above him. .

Even Aristotle pushes outside of man what he finds present only within man. The tendency of Greek thinking from Thales to Aristotle is to think that what is encountered within the inner life of man is an independent being existing for itself and to trace the things of the world back to this being.

Man's knowledge must pay the consequences when he thinks that the mediating of spirit with nature, which he himself is meant to accomplish, is accomplished by outer powers. He should immerse himself in his own inner being and seek there the point of connection between the sense world and the ideal world. If, instead of this, he looks into the outer world to find this point, then, because he cannot find it there, he must necessarily arrive eventually at the doubt in any reconciliation between the two powers. The period of Greek thought that follows Aristotle presents us with this stage of doubt. It announces itself with the Stoics and Epicureans and reaches its high-point with the Skeptics.

The Stoics and Epicureans feel instinctively that one cannot find the essential being of things along the path taken by their predecessors. They leave this path without bothering very much about finding a new one. For the older philosophers, the main thing was the world as a whole. They wanted to discover the laws of the world and believed that knowledge of man must result all by itself from knowledge of the world, because for them man was a part of the world-whole like all other things. The Stoics and Epicureans made man the main object of their reflections. They wanted to give his life its appropriate content. They thought about how man should live his life. Everything else was only a means to this end. The Stoics considered all philosophy to be worthwhile only to the extent that through it man could know how he is to live his life. They considered the right life for man to be one that is in harmony with nature. In order to realize this harmony with nature in one's own actions, one must first know what is in harmony with nature.

In the Stoics' teachings there lies an important admission about the human personality. Namely, that the human personality can be its own purpose and goal and that everything else, even knowledge, is there only for the sake of this personality.

The Epicureans went even further in this direction. Their striving consisted in shaping life in such a way that man would feel as content as possible in it or that it would afford him the greatest possible pleasure. One's own life stood so much in the foreground for them that they practiced knowledge only for the purpose of freeing man from superstitious fear and from the discomfort that befalls him when he does not understand nature.

A heightened human feeling of oneself runs through the views of the Stoics and Epicureans compared to those of older Greek thinkers.

This view appears in a finer, more spiritual way in the Skeptics. They said to themselves: When a person is forming ideas about things, he can form them only out of himself. And only out of himself can he draw the conviction that an idea corresponds to some thing. They saw nothing in the outer world that would provide a basis for connecting thing and idea. And they regarded as delusion and combated what anyone before them had said about any such bases.

The basic characteristic of the Skeptical view is modesty. Its adherents did not dare to deny that there is a connection in the outer world between idea and thing; they merely denied that man could know of any such connection. Therefore they did indeed make man the source of his knowing, but they did not regard this knowing as the expression of true wisdom.

Basically, Skepticism represents human knowing's declaration of bankruptcy. The human being succumbs to the preconception he has created for himself — that the truth is present outside him in a finished form — through the conviction he has gained that his truth is only an inner one, and therefore cannot be the right one at all.

Thales begins to reflect upon the world with utter confidence in the power of the human spirit. The doubt — that what human pondering must regard as the ground of the world could not actually be this ground — lay very far from his naive belief in man's cognitive ability. With the Skeptics a complete renunciation of real truth has taken the place of this belief.

The course of development taken by Greek thinking lies between the two extremes of naive, blissful confidence in man's cognitive ability and absolute lack of confidence in it. One can understand this course of development if one considers how man's mental pictures of the causes of the world have changed. What the oldest Greek philosophers thought these causes to be had sense-perceptible characteristics. Through this, one had a right to transfer these causes into the outer world. Like every other object in the sense world, the primal water of Thales belongs to outer reality. The matter became quite different when Parmenides stated that true existence lies in thinking. For, this thinking, in accordance with its true existence, is to be perceived only within man's inner being. Through Parmenides there first arose the great question: How does thought-existence, spiritual existence, relate to the outer existence that our senses perceive? One was accustomed then to picturing the relationship of the highest existence to that existence which surrounds us in daily life in the same way that Thales had thought the relationship to be between his sense-perceptible primal thing and the things that surround us. It is altogether possible to picture to oneself the emergence of all things out of the water that Thales presents as the primal source of all existence, to picture it as analogous to certain sense-perceptible processes that occur daily before our very eyes. And the urge to picture relations in the world surrounding us in the sense of such an analogy still remained even when, through Parmenides and his followers, pure thinking and its content, the world of ideas, were made into the primal source of all existence. Men were indeed ready to see that the spiritual world is a higher one than the sense world, that the deepest world-content reveals itself within the inner being of man, but they were not ready at the same time to picture the relationship between the sense world and the ideal world as an ideal one. They pictured it as a sense-perceptible relationship, as a factual emergence. If they had thought of it as spiritual, then they could peacefully have acknowledged that the content of the world of ideas is present only in the inner being of man. For then what is higher would not need to precede in time what is derivative. A sense-perceptible thing can reveal a spiritual content, but this content can first be born out of the sense-perceptible thing at the moment of revelation. This content is a later product of evolution than the sense world. But if one pictures the relationship to be one of emergence, then that from which the other emerges must also precede it in time. In this way the child — the spiritual world born of the sense world — was made into the mother of the sense world. This is the psychological reason why the human being transfers his world out into outer reality and declares — with reference to this his possession and product — that it has an objective existence in and for itself, and that he has to subordinate himself to it, or, as the case may be, that he can take possession of it only through revelation or in some other way by which the already finished truth can make its entry into his inner being.

This interpretation which man gives to his striving for truth, to his activity of knowing, corresponds with a profound inclination of his nature. Goethe characterized this inclination in his Aphorisms in Prose in the following words: "The human being never realizes just how anthropomorphic he is." And: "Fall and propulsion. To want to declare the movement of the heavenly bodies by these is actually a hidden anthropomorphism; it is the way a walker goes across a field. The lifted foot sinks down, the foot left behind strives forward and falls; and so on continuously from departing until arriving." All explanation of nature, indeed, consists in the fact that experiences man has of himself are interpreted into the object. Even the simplest phenomena are explained in this way. When we explain the propulsion of one body by another, we do so by picturing to ourselves that the one body exerts upon the other the same effect as we do when we propel a body. In the same way as we do this with something trivial, the religious person does it with his picture of God. He takes human ways of thinking and acting and interprets them into nature; and the philosophers we have presented, from Parmenides to Aristotle, also interpreted human thought-processes into nature. Max Stirner has this human need in mind when he says: "What haunts the universe and carries on its mysterious, 'incomprehensible' doings is, in fact, the arcane ghost that we call the highest being. And fathoming this ghost, understanding it, discovering reality in it (proving the 'existence of God') — this is the task men have set themselves for thousands of years; they tormented themselves with the horrible impossibility, with the endless work of the Danaides, of transforming the ghost into a nonghost, the unreal into a real, the spirit into a whole and embodied person. Behind the existing world they sought the 'thing-in-itself,' the essential being; they sought the non-thing behind the thing."

The last phase of Greek philosophy, Neo-Platonism, offers a splendid proof of how inclined the human spirit is to misconstrue its own being and therefore its relationship to the world. This teaching, whose most significant proponent is Plotin, broke with the tendency to transfer the content of the human spirit into a realm outside the living reality within which man himself stands. The Neo-Platonist seeks within his own soul the place at which the highest object of knowledge is to be found. Through that intensification of cognitive forces which one calls ecstasy, he seeks within himself to behold the essential being of world phenomena. The heightening of the inner powers of perception is meant to lift the human spirit onto a level of life at which he feels directly the revelation of this essential being. This teaching is a kind of mysticism. It is based on a truth that is to be found in every kind of mysticism. Immersion into one's own inner being yields the deepest human wisdom. But man must first prepare himself for this immersion. He must accustom himself to behold a reality that is free of everything the senses communicate to us. People who have brought their powers of knowledge to this height speak of an inner light that has dawned for them. Jakob Böhme, the Christian mystic of the seventeenth century, regarded himself as inwardly illumined in this way. He sees within himself the realm he must designate as the highest one knowable to man. He says: "Within the human heart (Gemüt) there lie the indications (Signatur), quite artfully set forth, of the being of all being."

Neo-Platonism sets the contemplation of the human inner world in the place of speculation about an outer world in the beyond. As a result, the highly characteristic phenomenon appears that the Neo-Platonist regards his own inner being as something foreign. One has taken things all the way to knowledge of the place at which the ultimate part of the world is to be sought; but one has wrongly interpreted what is to be found in this place. The Neo-Platonist therefore describes the inner experiences of his ecstasy like Plato describes the being of his supersensible world.

It is characteristic that Neo-Platonism excludes from the essential being of the inner world precisely that which constitutes its actual core. The state of ecstasy is supposed to occur only when self-consciousness is silent. It was therefore only natural that in Neo-Platonism the human spirit could not behold itself, its own being, in its true light.

The courses taken by the ideas that form the content of Greek philosophy found their conclusion in this view. They represent the longing of man to recognize, to behold, and to worship his own essential being as something foreign.

In the normal course of development within the spiritual evolution of the West, the discovery of egoism would have to have followed upon Neo-Platonism. That means, man would have to have recognized as his own being what he had considered to be a foreign being. He would have to have said to himself: The highest thing there is in the world given to man is his individual "I" whose being comes to manifestation within the inner life of the personality.

This natural course of Western spiritual development was held up by the spread of Christian teachings. Christianity presents, in popular pictures that are almost tangible, what Greek philosophy expressed in the language of sages. When one considers how deeply rooted in human nature the urge is to renounce one's own being, it seems understandable that this teaching has gained such incomparable power over human hearts. A high level of spiritual development is needed to satisfy this urge in a philosophical way. The most naive heart suffices to satisfy this urge in the form of Christian faith. Christianity does not present — as the highest being of the world — a finely spiritual content like Plato's world of ideas, nor an experience streaming forth from an inner light which must first be kindled; instead, it presents processes with attributes of reality that can be grasped by the senses. It goes so far, in fact, as to revere the highest being in a single historical person. The philosophical spirit of Greece could not present us with such palpable mental pictures. Such mental pictures lay in its past, in its folk mythology. Hamann, Herder's predecessor in the realm of theology, commented one time that Plato had never been a philosopher for children. But that it was for childish spirits that "the holy spirit had had the ambition to become a writer."

And for centuries this childish form of human self-estrangement has had the greatest conceivable influence upon the philosophical development of thought. Like fog the Christian teachings have hung before the light from which knowledge of man's own being should have gone forth. Through all kinds of philosophical concepts, the church fathers of the first Christian centuries seek to give a form to their popular mental pictures that would make them acceptable also to an educated consciousness. And the later teachers in the church, of whom Saint Augustine is the most significant, continue these efforts in the same spirit. The content of Christian faith had such a fascinating effect that there could be no question of doubt as to its truth, but only of lifting up of this truth into a more spiritual, more ideal sphere. The philosophy of the teachers within the church is a transforming of the content of Christian faith into an edifice of ideas. The general character of this thought-edifice could therefore be no other than that of Christianity: the transferring of man's being out into the world, self-renunciation. Thus it came about that Augustine again arrives at the right place, where the essential being of the world is to be found, and that he again finds something foreign in this place. Within man's own being he seeks the source of all truth; he declares the inner experiences of the soul to be the foundations of knowledge. But the teachings of Christian faith have set an extra-human content at the place where he was seeking. Therefore, at the right place, he found the wrong beings.

There now follows a centuries-long exertion of human thinking whose sole purpose, by expending all the power of the human spirit, was to bring proof that the content of this spirit is not to be sought within this spirit but rather at that place to which Christian faith has transferred this content. The movement in thought that grew up out of these efforts is called Scholasticism. All the hair-splittings of the Schoolmen can be of no interest in the context of the present essay. For that movement in ideas does not represent in the least a development in the direction of knowledge of the personal "I."

The thickness of the fog in which Christianity enshrouded human self-knowledge becomes most evident through the fact that the Western spirit, out of itself, could not take even one step on the path to this self-knowledge. The Western spirit needed a decisive push from outside. It could not find upon the ground of the soul what it had sought so long in the outer world. But it was presented with proof that this outer world could not be constituted in such a way that the human spirit could find there the essential being it sought. This push was given by the blossoming of the natural sciences in the sixteenth century. As long as man had only an imperfect picture of how natural processes are constituted, there was room in the outer world for divine beings and for the working of a personal divine will. But there was no longer a place, in the natural picture of the world sketched out by Copernicus and Kepler, for the Christian picture. And as Galileo laid the foundations for an explanation of natural processes through natural laws, the belief in divine laws had to be shaken.

Now one had to seek in a new way the being that man recognizes as the highest and that had been pushed out of the external world for him.

Francis Bacon drew the philosophical conclusions from the presuppositions given by Copernicus, Kepler, and Galileo. His service to the Western world view is basically a negative one. He called upon man in a powerful way to direct his gaze freely and without bias upon reality, upon life. As obvious as this call seems, there is no denying that the development of Western thought has sinned heavily against it for centuries. Man's own "I" also belongs within the category of real things. And does it not almost seem as though man's natural predisposition makes him unable to look at this "I" without bias? Only the development of a completely unbiased sense, directed immediately upon what is real, can lead to self-knowledge. The path of knowledge of nature is also the path of knowledge of the "I."

Two streams now entered into the development of Western thought that tended, by different paths, in the direction of the new goals of knowledge necessitated by the natural sciences. One goes back to Jakob Böhme, the other to René Descartes.

Jakob Böhme and Descartes no longer stood under the influence of Scholasticism. Böhme saw that nowhere in cosmic space was there a place for heaven; he therefore became a mystic. He sought heaven within the inner being of man. Descartes recognized that the adherence of the Schoolmen to Christian teachings was only a matter of centuries-long habituation to these pictures. Therefore he considered it necessary first of all to doubt these habitual pictures and to seek a way of knowledge by which man can arrive at a kind of knowing whose certainty he does not assert out of habit, but which can be guaranteed at every, moment through his own spiritual powers.

Those are therefore strong initial steps which — both with Böhme and with Descartes — the human "I" takes to know itself. Both were nevertheless overpowered by the old preconceptions in what they brought forth later. It has already been indicated that Jakob Böhme has a certain spiritual kinship with the Neo-Platonists. His knowledge is an entering into his own inner being. But what confronts him within this inner being is not the "I" of man but rather only the Christian God again. He becomes aware that within his own heart (Gemüt) there lies what the person who needs knowledge is craving. Fulfillment of the greatest human longings streams toward him from there. But this does not lead him to the view that the "I," by intensifying its cognitive powers, is also able out of itself to satisfy its demands. This brings him, rather, to the belief that, on the path of knowledge into the human heart, he had truly found the God whom Christianity had sought upon a false path. Instead of self-knowledge, Jakob Böhme seeks union with God; instead of life with the treasures of his own inner being, he seeks a life in God.

It is obvious that the way man thinks about his actions, about his moral life, will also depend upon human self-knowledge or self-misapprehension. The realm of morality does in fact establish itself as a kind of upper story above the purely natural processes. Christian belief, which already regards these natural processes as flowing from the divine will, seeks this will all the more within morality. Christian moral teachings show more clearly than almost anything else the distortedness of this world view. No matter how enormous the sophistry is that theology has applied to this realm: questions remain which, from the standpoint of Christianity, show definite features of considerable contradiction. If a primal being like the Christian God is assumed, it is incomprehensible how the sphere of human action can fall into two realms: into that of the good and into that of the evil. For, all human actions would have to flow from the primal being and consequently bear traits homogeneous with their origin. Human actions would in fact have to be divine. Just as little can human responsibility be explained on this basis. Man is after all directed by the divine will. He can therefore give himself up only to this will; he can let happen through him only what God brings about.

In the views one held about morality, precisely the same thing occurred as in one's views about knowledge. Man followed his inclination to tear his own self out of himself and to set it up as something foreign. And just as in the realm of knowledge no other content could be given to the primal being — regarded as lying outside man — than the content drawn from his own inner being, so no moral aims and impulses for action could be found in this primal being except those belonging to the human soul. What man, in his deepest inner being, was convinced should happen, this he regarded as something willed by the primal being of the world. In this way a duality in the ethical realm was created. Over against the self that one had within oneself and out of which one had to act, one set one's own content as something morally determinative. And through this, moral demands could arise. Man's self was not allowed to follow itself; it had to follow something foreign. Selflessness in one's actions in the moral field corresponds to self-estrangement in the realm of knowledge. Those actions are good in which the "I" follows something foreign; those actions are bad, on the other hand, in which it follows itself. In self-will Christianity sees the source of all evil. That could never have happened if one had seen that everything moral can draw its content only out of one's own self. One can sum up all the Christian moral teachings in one sentence:

If man admits to himself that he can follow only the commandments of his own being and if he acts according to them, then he is evil; if this truth is hidden from him and if he sets — or allows to be set — his own commandments as foreign ones over himself in order to act according to them, then he is good.

The moral teaching of selflessness is elaborated perhaps more completely than anywhere else in a book from the fourteenth century, German Theology. The author of this book is unknown to us. He carried self-renunciation far enough to be sure that his name did not come down to posterity. In this book it is stated: "That is no true being and has no being which does not exist within the perfect; rather it is by chance or it is a radiance and a shining that is no being or has no being except in the fire from which the radiance flows, or in the sun, or in the light. The Bible speaks of faith and the truth: sin is nothing other than the fact that the creature turns himself away from the unchangeable good and toward the changeable good, which means that he turns from the perfect to the divided and to the imperfect and most of all to himself. Now mark. If the creature assumes something good — such as being, living, knowing, recognizing, capability, and everything in short that one should call good — and believes that he is this good, or that it is his or belongs to him, or that it is of him, no matter how often nor how much results from this, then he is going astray. What else did the devil do or what else was his fall and estrangement than that he assumed that he was also something and something would be his and something would also belong to him? That assumption and his "I" and his "me," his "for me" and his "mine," that was his estrangement and his fall. That is how it still is. For, everything that one considers good or should call good belongs to no one, but only to the eternal true good which God is alone, and whoever assumes it of himself acts wrongly and against God."

A change in moral views from the old Christian ones is also connected with the turn that Jakob Böhme gave to man's relationship to God. God still works as something higher in the human soul to effect the good, but He does at least work within this self and not from outside upon the self. An internalizing of moral action occurs thereby. The rest of Christianity demanded only an outer obedience to the divine will. With Jakob Böhme the previously separated entities — the really personal and the personal that was made into God — enter into a living relationship. Through this, the source of the moral is indeed now transferred into man's inner being, but the moral principle of selflessness seems to be even more strongly emphasized. If God is regarded as an outer power, then the human self is the one actually acting. It acts either in God's sense or against it. But if God is transferred into man's inner being, then man himself no longer acts, but rather God in him. God expresses himself directly in human life. Man foregoes any life of his own; he makes himself a part of the divine life. He feels himself in God, God in himself; he grows into the primal being; he becomes an organ of it.

In this German mysticism man has therefore paid for his participation in the divine life with the most complete extinguishing of his personality, of his "I." Jakob Böhme and the mystics who were of his view did not feel the loss of the personal element. On the contrary: they experienced something particularly uplifting in the thought that they were directly participating in the divine life, that they were members in a divine organism. An organism cannot exist, after all, without its members. The mystic therefore felt himself to be something necessary within the world-whole, as a being that is indispensable to God. Angelus Silesius, the mystic who felt things in the same spirit as Jakob Böhme, expresses this in a beautiful statement:

I know that without me God cannot live an instant, Came I to naught, he needs must yield the spirit.

And even more characteristically in another one:

Without me God cannot a single worm create; Do I not co-maintain it, it must at once crack open.

The human "I" asserts its rights here in the most powerful way vis-à-vis its own image which it has transferred into the outer world. To be sure, the supposed primal being is not yet told that it is man's own being set over against himself, but at least man's own being is considered to be the maintainer of the divine primal ground.

Descartes had a strong feeling for the fact that man, through his thought-development, had brought himself into a warped relationship with the world. Therefore, to begin with, he met everything that had come forth from this thought-development with doubt. Only when one doubts everything that the centuries have developed as truths can one — in his opinion — gain the necessary objectivity for a new point of departure. It lay in the nature of things that this doubt would lead Descartes to the human "I." For, the more a person regards everything else as something that he still must seek, the more he will have an intense feeling of his own seeking personality. He can say to himself: Perhaps I am erring on the paths of existence; then the erring one is thrown all the more clearly back upon himself. Descartes' Cogito, ergo sum (I think, therefore I am) indicates this. Descartes presses even further. He is aware that the way man arrives at knowledge of himself should be a model for any other knowledge he means to acquire. Clarity and definiteness seem to Descartes to be the most prominent characteristics of self-knowledge. Therefore he also demands these two characteristics of all other knowledge. Whatever man can distinguish just as clearly and definitely as his own existence: only that can stand as certain.

With this, the absolutely central place of the "I" in the world-whole is at least recognized in the area of cognitive methodology. Man determines the how of his knowledge of the world according to the how of his knowledge of himself, and no longer asks for any outer being to justify this how. Man does not want to think in the way a god prescribes knowing activity to be, but rather in the way he determines this for himself. From now on, with respect to the world, man draws the power of his wisdom from himself.

In connection with the what, Descartes did not take the same step. He set to work to gain mental pictures about the world, and — in accordance with the cognitive principle just presented — searched through his own inner being for such mental pictures. There he found the mental picture of God. It was of course nothing more than the mental picture of the human "I." But Descartes did not recognize this. The idea of God as the altogether most perfect being » brought his thinking onto a completely wrong path. This one characteristic, that of the altogether greatest perfection, outshone for him all the other characteristics of the central being. He said to himself: Man, who is himself imperfect, cannot out of himself create the mental picture of an altogether most perfect being. Consequently this altogether most perfect being exists. If Descartes had investigated the true content of his mental picture of God, he would have found that it is exactly the same as the mental picture of the "I," and that perfection is only a conceptual enhancement of this content. The essential content of an ivory ball is not changed by my thinking of it as infinitely large. Just as little does the mental picture of the "I" become something else through such an enhancement.

The proof that Descartes brings for the existence of God is therefore again nothing other than a paraphrasing of the human need to make one's own "I," in the form of a being outside man, into the ground of the world. But here indeed the fact presents itself with full clarity that man can find no content of its own for this primal being existing outside man, but rather can only lend this being the content of his mental picture of the "I" in a form that has not been significantly changed.

Spinoza took no step forward on the path that must lead to the conquest of the mental picture of the "I"; he took a step backward. For Spinoza has no feeling of the unique position of the human "I." For him the stream of world processes consists only in a system of natural necessity, just as for the Christian philosophers it consisted only in a system of divine acts of will. Here as there the human "I" is only a part within this system. For the Christian, man is in the hands of God; for Spinoza he is in those of natural world happenings. With Spinoza the Christian God received a different character. A philosopher who has grown up in a time when natural-scientific insights are blooming cannot acknowledge a God who directs the world arbitrarily; he can acknowledge only a primal being who exists because his existence, through itself, is a necessity, and who guides the course of the world according to the unchangeable laws that flow from his own absolutely necessary being. Spinoza has no consciousness of the fact that man takes the image in which he pictures this necessity from his own content. For this reason Spinoza's moral ideal also becomes something impersonal, unindividual. In accordance with his presuppositions he cannot indeed see his ideal to be in the perfecting of the "I," in the enhancement of man's own powers, but rather in the permeating of the "I" with the divine world content, with the highest knowledge of the objective God. To lose oneself in this God should be the goal of human striving.

The path Descartes took — to start with the "I" and press forward to world knowledge — is extended from now on by the philosophers of modern times. The Christian theological method, which had no confidence in the power of the human "I" as an organ of knowledge, at least was overcome. One thing was recognized: that the "I" itself must find the highest being. The path from there to the other point — to the insight that the content lying within the "I" is also the highest being — is, to be sure, a long one.

Less thoughtfully than Descartes did the two English philosophers Locke and Hume approach their investigation of the paths that the human "I" takes to arrive at enlightenment about itself and the world. One thing above all was lacking in both of them: a healthy, free gaze into man's inner being. Therefore they could also gain no mental picture of the great difference that exists between knowledge of outer things and knowledge of the human "I." Everything they say relates only to the acquisition of outer knowledge. Locke entirely overlooks the fact that man, by enlightening himself about outer things, sheds a light upon them that streams from his own inner being. He believes therefore that all knowledge stems from experience. But what is experience? Galileo sees a swinging church lamp. It leads him to find the laws by which a body swings. He has experienced two things: firstly, through his senses, outer processes; secondly, from out of himself, the mental picture of a law that enlightens him about these processes, that makes them comprehensible. One can now of course call both of these experience. But then one fails to recognize the difference, in fact, that exists between the two parts of this cognitive process. A being that could not draw upon the content of his being could stand eternally before the swinging church lamp: the sense perception would never complement itself with a conceptual law. Locke and all who think like him allow themselves to be deceived by something — namely by the way the content of what is to be known approaches us. It simply rises up, in fact, upon the horizon of our consciousness. Experience consists in what thus arises. But the fact must be recognized that the content of the laws of experience is developed by the "I" in its encounter with experience. Two things reveal themselves in Hume. One is that, as already mentioned, he does not recognize the nature of the "I," and therefore, exactly like Locke, derives the content of the laws from experience. The other thing is that this content, by being separated from the "I," loses itself completely in indefiniteness, hangs freely in the air without support or foundation. Hume recognizes that outer experience communicates only unconnected processes, that it does not at the same time, along with these processes, provide the laws by which they are connected. Since Hume knows nothing about the being of the "I," he also cannot derive from it any justification for connecting the processes. He therefore derives these laws from the vaguest source one could possibly imagine: from habit. A person sees that a certain process always follows upon another; the fall of a stone is followed by the indentation of the ground on which it falls. As a result man habituates himself to thinking of such processes as connected. All knowledge loses its significance if one takes one's start from such presuppositions. The connection between the processes and their laws acquires something of a purely chance nature.

We see in George Berkeley a person for whom the creative being of the "I" has come fully to consciousness. He had a clear picture of the "I's" own activity in the coming about of all knowledge. When I see an object, he said to himself, I am active. I create my perception for myself. The object of my perception would remain forever beyond my consciousness, it would not be there for me, if I did not continuously enliven its dead existence by my activity. I perceive only my enlivening activity, and not what precedes it objectively as the dead thing. No matter where I look within the sphere of my consciousness: everywhere I see myself as the active one, as the creative one. In Berkeley's thinking, the "I" acquires a universal life. What do I know of any existence of things, if I do not picture this existence?

For Berkeley the world consists of creative spirits who out of themselves form a world. But at this level of knowledge there again appeared, even with him, the old preconception. He indeed lets the "I" create its world for itself, but he does not give it at the same time the power to create itself out of itself. It must again proffer a mental picture of God. The creative principle in the "I" is God, even for Berkeley.

But this philosopher does show us one thing. Whoever really immerses himself into the essential being of the creative "I" does not come back out of it again to an outer being except by forcible means. And Berkeley does proceed forcibly. Under no compelling necessity he traces the creativity of the "I" back to God. Earlier philosophers emptied the "I" of its content and through this gained a content for their God. Berkeley does not do this. Therefore he can do nothing other than set, beside the creative spirits, yet one more particular spirit that basically is of exactly the same kind as they and therefore completely unnecessary, after all.

This is even more striking in the German philosopher Leibniz. He also recognized the creative activity of the "I." He had a very clear overview of the scope of this activity; he saw that it was inwardly consistent, that it was founded upon itself. The "I" therefore became for him a world in itself, a monad. And everything that has existence can have it only through the fact that it gives itself a self-enclosed content. Only monads, i.e., beings creating out of and within themselves, exist: separate worlds in themselves that do not have to rely on anything outside themselves. Worlds exist, no world. Each person is a world, a monad, in himself. If now these worlds are after all in accord with one another, if they know of each other and think the contents of their knowledge, then this can only stem from the fact that a predestined accord (pre-established harmony) exists. The world, in fact, is arranged in such a way that the one monad creates out of itself something which corresponds to the activity in the others. To bring about this accord Leibniz of course again needs the old God. He has recognized that the "I" is active, creative, within his inner being, that it gives its content to itself; the fact that the "I" itself also brings this content into relationship with the other content of the world remained hidden to him. Therefore he did not free himself from the mental picture of God. Of the two demands that lie in the Goethean statement — "If I know my relationship to myself and to the outer world, then I call it truth" — Leibniz understood only the one.

This development of European thought manifests a very definite character. Man must draw out of himself the best that he can know. He in fact practices self-knowledge. But he always shrinks back again from the thought of also recognizing that what he has created is in fact self-created. He feels himself to be too weak to carry the world. Therefore he saddles someone else with this burden. And the goals he sets for himself would lose their weight for him if he acknowledged their origin to himself; therefore he burdens his goals with powers that he believes he takes from outside. Man glorifies his child but without wanting to acknowledge his own fatherhood.

In spite of the currents opposing it, human self-knowledge made steady progress. At the point where this self-knowledge began to threaten man's belief in the beyond, it met Kant. Insight into the nature of human knowing had shaken the power of those proofs which people had thought up to support belief in the beyond. One had gradually gained a picture of real knowledge and therefore saw through the artificiality and tortured nature of the seeming ideas that were supposed to give enlightenment about other-worldly powers. A devout, believing man like Kant could fear that a further development along this path would lead to the disintegration of all faith. This must have seemed to his deeply religious sense like a great, impending misfortune for mankind. Out of his fear of the destruction of religious mental pictures there arose for him the need to investigate thoroughly the relationship of human knowing to matters of faith. How is knowing possible and over what can it extend itself? That is the question Kant posed himself, with the hope, right from the beginning, of being able to gain from his answer the firmest possible support for faith.

Kant took up two things from his predecessors. Firstly, that there is a knowledge in some areas that is indubitable. The truths of pure mathematics and the general teachings of logic and physics seem to him to be in this category. Secondly, he based himself upon Hume in his assertion that no absolutely sure truths can come from experience. Experience teaches only that we have so and so often observed certain connections; nothing can be determined by experience as to whether these connections are also necessary ones. If there are indubitable, necessary truths and if they cannot stem from experience: then from what do they stem? They must be present in the human soul before experience. Now it becomes a matter of distinguishing between the part of knowledge that stems from experience and the part that cannot be drawn from this source of knowledge. Experience occurs through the fact that I receive impressions. These impressions are given through sensations. The content of these sensations cannot be given us in any other way than through experience. But these sensations, such as light, color, tone, warmth, hardness, etc., would present only a chaotic tangle if they were not brought into certain interconnections. In these interconnections the contents of sensation first constitute the objects of experience. An object is composed of a definitely ordered group of the contents of sensation. In Kant's opinion, the human soul accomplishes the ordering of these contents of sensation into groups. Within the human soul there are certain principles present by which the manifoldness of sensations is brought into objective unities. Such principles are space, time, and certain connections such as cause and effect. The contents of sensation are given me, but not their spatial interrelationships nor temporal sequence. Man first brings these to the contents of sensation. One content of sensation is given and another one also, but not the fact that one is the cause of the other. The intellect first makes this connection. Thus there lie within the human soul, ready once and for all, the ways in which the contents of sensation can be connected. Thus, even though we can take possession of the contents of sensation only through experience, we can, nevertheless, before all experience, set up laws as to how these contents of sensation are to be connected. For, these laws are the ones given us within our own souls.

We have, therefore, necessary kinds of knowledge. But these do not relate to a content, but only to ways of connecting contents. In Kant's opinion, we will therefore never draw knowledge with any content out of the human soul's own laws. The content must come through experience. But the otherworldly objects of faith can never become the object of any experience. Therefore they also cannot be attained through our necessary knowledge. We have a knowledge from experience and another, necessary, experience-free knowledge as to how the contents of experience can be connected. But we have no knowledge that goes beyond experience. The world of objects surrounding us is as it must be in accordance with the laws of connection lying ready in our soul. Aside from these laws we do not know how this world is "in-itself." The world to which our knowledge relates itself is no such "in-itselfness" but rather is an appearance for us.

Obvious objections to these Kantian views force themselves upon the unbiased person. The difference in principle between the particulars (the contents of sensation) and the way of connecting these particulars does not consist, with respect to knowledge, in the way we connect things as Kant assumes it to. Even though one element presents itself to us from outside and the other comes forth from our inner being, both elements of knowledge nevertheless form an undivided unity. Only the abstracting intellect can separate light, warmth, hardness, etc., from spatial order, causal relationship, etc. In reality, they document, with respect to every single object, their necessary belonging together. Even the designation of the one element as "content" in contrast to the other element as a merely "connecting" principle is all warped. In truth, the knowledge that something is the cause of something else is a knowledge with just as much content as the knowledge that it is yellow. If the object is composed of two elements, one of which is given from outside and the other from within, it follows that, for our knowing activity, elements which actually belong together are communicated along two different paths. It does not follow, however, that we are dealing with two things that are different from each other and that are artificially coupled together.

Only by forcibly separating what belongs together can Kant therefore support his view. The belonging together of the two elements is most striking in knowledge of the human "I." Here one element does not come from outside and the other from within; both arise from within. And here both are not only one content but also one completely homogeneous content.

What mattered to Kant — his heart's wish that guided his thoughts far more than any unbiased observation of the real factors — was to rescue the teachings relative to the beyond. What knowledge had brought about as support for these teachings in the course of long ages had decayed. Kant believed he had now shown that it is anyway not for knowledge to support such teachings, because knowledge has to rely on experience, and the things of faith in the beyond cannot become the object of any experience. Kant believed he had thereby created a free space where knowledge could not get in his way and disrupt him as he built up there a faith in the beyond. And he demands, as a support for moral life, that one believe in the things in the beyond. Out of that realm from which no knowledge comes to us, there sounds the despotic voice of the categorical imperative which demands of us that we do the good. And in order to establish a moral realm we would in fact need all that about which knowledge can tell us nothing. Kant believed he had achieved what he wanted: "I therefore had to set knowledge aside in order to make room for faith."

The great philosopher in the development of Western thought who set out in direct pursuit of a knowledge of human self-awareness is Johann Gottlieb Fichte. It is characteristic of him that he approaches this knowledge without any presuppositions, with complete lack of bias. He has the clear, sharp awareness of the fact that nowhere in the world is a being to be found from which the "I" could be derived. It can therefore be derived only from itself. Nowhere is a power to be found from which the existence of the "I" flows. Everything the "I" needs, it can acquire only out of itself. Not only does it gain enlightenment about its own being through self-observation; it first posits this being into itself through an absolute, unconditional act. "The 'I' posits itself, and it is by virtue of this mere positing of itself; and conversely: The 'I' is, and posits its existence, by virtue of its mere existence. It is at the same time the one acting and the product of its action; the active one and what is brought forth by the activity; action and deed are one and the same; and therefore the 'I am' is the expression of an active deed." Completely undisturbed by the fact that earlier philosophers have transferred the entity he is describing outside man, Fichte looks at the "I" naively. Therefore the "I" naturally becomes for him the highest being. "That whose existence (being) merely consists in the fact that it posits itself as existing is the 'I' as absolute subject. In the way that it posits itself, it is, and in the way that it is, it posits itself: and the 'I' exists accordingly for the 'I,' simply and necessarily. What does not exist for itself is no 'I' ... One certainly hears the question raised: What was I anyway, before I came to self-awareness? The obvious answer to that is: I was not at all; for I was not I... To posit oneself and to be are, for the 'I,' completely the same." The complete, bright clarity about one's own "I," the unreserved illumination of one's personal, human entity, becomes thereby the starting point of human thinking. The result of this must be that man, starting here, sets out to conquer the world. The second of the Goethean demands mentioned above, knowledge of my relationship to the world, follows upon the first — knowledge of the relationship that the "I" has to itself. This philosophy, built upon self-knowledge, will speak about both these relationships, and not about the derivation of the world from some primal being. One could now ask: Is man then supposed to set his own being in place of the primal being into which he transferred the world origins? Can man then actually make himself the starting point of the world? With respect to this it must be emphasized that this question as to the world origins stems from a lower sphere. In the sequence of the processes given us by reality, we seek the causes for the events, and then seek still other causes for the causes, and soon. We are now stretching the concept of causation. We are seeking a final cause for the whole world. And in this way the concept of the first, absolute primal being, necessary in itself, fuses for us with the idea of the world cause. But that is a mere conceptual construction. When man sets up such conceptual constructions, they do not necessarily have any justification. The concept of a flying dragon also has none. Fichte takes his start from the "I" as the primal being, and arrives at ideas that present the relationship of this primal being to the rest of the world in an unbiased way, but not under the guise of cause and effect. Starting from the "I," Fichte now seeks to gain ideas for grasping the rest of the world. Whoever does not want to deceive himself about the nature of what one can call cognition or knowledge can proceed in no other way. Everything that man can say about the being of things is derived from the experiences of his inner being. "The human being never realizes just how anthropomorphic he is." (Goethe) In the » explanation of the simplest phenomena, in the propulsion of one body by another, for example, there lies an anthropomorphism. The conclusion that the one body propels the other is already anthropomorphic. For, if one wants to go beyond what the senses tell us about the occurrence, one must transfer onto it the experience our body has when it sets a body in the outer world into motion. We transfer our experience of propelling something onto the occurrence in the outer world, and also speak there of propulsion when we roll one ball and as a result see a second ball go rolling. For we can observe only the movements of the two balls, and then in addition think the propulsion in the sense of our own experiences. All physical explanations are anthropomorphisms, attributing human characteristics to nature. But of course it does not follow from this what has so often been concluded from this: that these explanations have no objective significance for the things. A part of the objective content lying within the things, in fact, first appears when we shed that light upon it which we perceive in our own inner being.

Whoever, in Fichte's sense, bases the being of the "I" entirely upon itself can also find the sources of moral action only within the "I" alone. The "I" cannot seek harmony with some other being, but only with itself. It does not allow its destiny to be prescribed, but rather gives any such destiny to itself. Act according to the basic principle that you can regard your actions as the most worthwhile possible. That is about how one would have to express the highest principle of Fichte's moral teachings. "The essential character of the 'I,' in which it distinguishes itself from everything that is outside it, consists in a tendency toward self-activity for the sake of self-activity; and it is this tendency that is thought when the 'I,' in and for itself, without any relationship to something outside it, is thought." An action therefore stands on an ever higher level of moral value, the more purely it flows from the self-activity and self-determination of the "I."

In his later life Fichte changed his self-reliant, absolute "I" back into an external God again; he therefore sacrificed true self-knowledge, toward which he had taken so many important steps, to that self-renunciation which stems from human weakness. The last books of Fichte are therefore of no significance for the progress of this self-knowledge.

The philosophical writings of Schiller, however, are important for this progress. Whereas Fichte expressed the self-reliant independence of the "I" as a general philosophical truth, Schiller was more concerned with answering the question as to how the particular "I" of the simple human individuality could live out this self-activity in the best way within itself.

Kant had expressly demanded the suppression of pleasure as a pre-condition for moral activity. Man should not carry out what brings him satisfaction; but rather what the categorical imperative demands of him. According to his view an action is all the more moral the more it is accomplished with the quelling of all feeling of pleasure, out of mere heed to strict moral law. For Schiller this diminishes human worth. Is man in his desire for pleasure really such a low being that he must first extinguish this base nature of his in order to be virtuous? Schiller criticizes any such degradation of man in the satirical epigram (Xenie):

Gladly I serve all my friends, but do so alas out of liking; Therefore it rankles me often that I'm not a virtuous man.

No, says Schiller, human instincts are capable of such ennobling that it is a pleasure to do the good. The strict "ought to" transforms itself in the ennobled man into a free "wanting to." And someone who with pleasure accomplishes what is moral stands higher on the moral world scale than someone who must first do violence to his own being in order to obey the categorical imperative.

Schiller elaborated this view of his in his Letters on the Aesthetic Education of the Human Race. There hovers before him the picture of a free individuality who can calmly give himself over to his egoistical drives because these drives, out of themselves, want what can be accomplished by the unfree, ignoble personality only when it suppresses its own needs. The human being, as Schiller expressed it, can be unfree in two respects: firstly, if he is able to follow only his blind, lower instincts. Then he acts out of necessity. His drives compel him; he is not free. Secondly, however, that person also acts unfreely who follows only his reason. For, reason sets up principles of behavior according to logical rules. A person who merely follows reason acts unfreely because he subjugates himself to logical necessity. Only that person acts freely out of himself for whom what is reasonable has united so deeply with his individuality , has gone over so fully into his flesh and blood, that he carries out with the greatest pleasure what someone standing morally less high can accomplish only through the most extreme self-renunciation and the strongest compulsion.

Friedrich Joseph Schelling wanted to extend the path Fichte had taken. Schelling took his start from the unbiased knowledge of the "I" that his predecessor had achieved. The "I" was recognized as a being that draws its existence out of itself. The next task was to bring nature into a relationship with this self-reliant "I." It is clear: If the "I" is not to transfer the actual higher being of things into the outer world again, then it must be shown that the "I," out of itself, also creates what we call the laws of nature. The structure of nature must therefore be the material system, outside in space, of what the "I," within its inner being, creates in a spiritual way. "Nature must be visible spirit, and spirit must be invisible nature. Here, therefore, in the absolute identity of the spirit in us and of nature outside of us, must the problem be solved as to how a nature outside of us is possible." "The outer world lies open before us, in order for us to find in it again the history of our spirit."

Schelling, therefore, sharply illuminates the process that the philosophers have interpreted wrongly for so long. He shows that out of one being the clarifying light must fall upon all the processes of the world; that the "I" can recognize one being in all happenings; but he no longer sets forth this being as something lying outside the "I"; he sees it within the "I." The "I" finally feels itself to be strong enough to enliven the content of world phenomena from out of itself. The way in which Schelling presented nature in detail as a material development out of the "I" does not need to be discussed here. The important thing in this essay is to show in what way the "I" has reconquered for itself the sphere of influence which, in the course of the development of Western thought, it had ceded to an entity that it had itself created. For this reason Schelling's other writings also do not need to be considered in this context. At best they add only details to the question we are examining. Exactly like Fichte, Schelling abandons clear self-knowledge again, and seeks then to trace the things flowing from the self back to other beings. The later teachings of both thinkers are reversions to views which they had completely overcome in an earlier period of life.

The philosophy of Georg Wilhelm Friedrich Hegel is a further bold attempt to explain the world on the basis of a content lying within the "I." Hegel sought, comprehensively and thoroughly, to investigate and present the whole content of what Fichte, in incomparable words to be sure, had characterized: the being of the human "I." For Hegel also regards this being as the actual primal thing, as the "in-itselfness of things." But Hegel does something peculiar. He divests the "I" of everything individual, personal. In spite of the fact that it is a genuine true "I" which Hegel takes as a basis for world phenomena, this "I" seems impersonal, unindividual, far from an intimate, familiar "I," almost like a god. In just such an unapproachable, strictly abstract form does Hegel, in his logic, expound upon the content of the in-itselfness of the world. The most personal thinking is presented here in the most impersonal way. According to Hegel, nature is nothing other than the content of the "I" that has been spread out in space and time. Nature is this ideal content in a different state. "Nature is spirit estranged from itself." Within the individual human spirit Hegel's stance toward the impersonal "I" is personal. Within self-consciousness, the being of the "I" is not an in-itself, it is also for-itself; the human spirit discovers that the highest world content is his own content.

Because Hegel seeks to grasp the being of the "I" at first impersonally, he also does not designate it as "I," but rather as idea. But Hegel's idea is nothing other than the content of the human "I" freed of all personal character. This abstracting of everything personal manifests most strongly in Hegel's views about the spiritual life, the moral life. It is not the single, personal, individual "I" of man that can decide its own destiny, but rather it is the great, objective, impersonal world "I," which is abstracted from man's individual "I"; it is the general world reason, the world idea. The individual "I" must submit to this abstraction drawn from its own being. The world idea has instilled the objective spirit into man's legal, state, and moral institutions, into the historical process. Relative to this objective spirit, the individual is inferior, coincidental. Hegel never tires of emphasizing again and again that the chance, individual "I" must incorporate itself into the general order, into the historical course of spiritual evolution. It is the despotism of the spirit over the bearer of this spirit that Hegel demands.

It is a strange last remnant of the old belief in God and in the beyond that still appears here in Hegel. All the attributes with which the human "I," turned into an outer ruler of the world, was once endowed have been dropped, and only the attribute of logical generality remains. The Hegelian world idea is the human "I," and Hegel's teachings recognize this expressly, for at the pinnacle of culture man arrives at the point, according to this teaching, of feeling his full identity with this world "I." In art, religion, and philosophy man seeks to incorporate into his particular existence what is most general; the individual spirit permeates itself with the general world reason. Hegel portrays the course of world history in the following way: "If we look at the destiny of world-historical individuals, they have had the good fortune to be the managing directors of a purpose that was one stage in the progress of the general spirit. One can call it a trick of world reason for it to use these human tools; for it allows them to carry out their own purposes with all the fury of their passion, and yet remains not only unharmed itself but even brings forth itself. The particular is usually too insignificant compared to the general: individuals are sacrificed and abandoned. World history thus presents itself as the battle of individuals, and in the field of this particularization, things take their completely natural course. Just as in animal nature the preservation of life is the purpose and instinct of the individual creature, and just as here, after all, reason, the general, predominates and the individuals fall, thus so do things in the spiritual world also take their course. The passions mutually destroy each other; only reason is awake, pursues its purpose, and prevails." But for Hegel, the highest level of development of human culture is also not presented in this sacrificing of the particular individuals to the good of general world reason, but rather in the complete interpenetration of the two. In art, religion, and philosophy, the individual works in such a way that his work is at the same time a content of the general world reason. With Hegel, through the factor of generality that he laid into the world "I," the subordination of the separate human "I" to this world "I" still remained.

Ludwig Feuerbach sought to put an end to this subordination by stating in powerful terms how man transfers the being of his "I" into the outer world in order then to place himself over against it, acknowledging, obeying, revering it as though it were a God. "God is the revealed inner being, the expressed self, of man; religion is the festive disclosing of the hidden treasures of man, the confessing of his innermost thoughts, the public declaration of his declarations of love." But even Feuerbach has not yet cleansed the idea of this "I" of the factor of generality. For him the general human "I" is something higher than the individual, single "I." And even though as a thinker he does not, like Hegel, objectify this general "I" into a cosmic being existing in itself, still, in the moral context, over against the single human being, he does set up the general concept of a generic man, and demands that the individual should raise himself above the limitations of his individuality.

Max Stirner, in his book The Individual and What Is His (Der Einzige und sein Eigentum), published in 1844, demanded of the "I" in a radical way that it finally recognize that all the beings it has set above itself in the course of time were cut by it from its own body and set up in the outer world as idols. Every god, every general world reason, is an image of the "I" and has no characteristics different from the human "I." And even the concept of the general "I" was extracted from the completely individual "I" of every single person.

Stirner calls upon man to throw off everything general about himself and to acknowledge to himself that he is an individual. "You are indeed more than a Jew, more than a Christian, etc., but you are also more than a man. Those are all ideas; you, however, are in the flesh. Do you really believe, therefore, that you can ever become 'man as such'?" "*I* am man! I do not first have to produce man in myself, because he already belongs to me as all my characteristics do." "Only I am not an abstraction alone; I am the all in all;... I am no mere thought, but I am at the same time full of thoughts, a thought-world. Hegel condemns what is one's own, what is mine ... 'Absolute thinking' is that thinking which forgets that it is my thinking, that I think, and that thinking exists only through me. As 'I,' however, I again swallow what is mine, am master over it; it is only my opinion that I can change at every moment, i.e., that I can destroy, that I can take back into myself and can devour." "The thought is only my own when I can indeed subjugate it, but it can never subjugate me, never fanaticize me and make me the tool of its realization." All the beings placed over the "I" finally shatter upon the knowledge that they have only been brought into the world by the "I." "The beginning of my thinking, namely, is not a thought, but rather I, and therefore I am also its goal, just as its whole course is then only the course of my self-enjoyment."

In Stirner's sense, one should not want to define the individual "I" by a thought, by an idea. For, ideas are something general; and through any such definition, the individual — at least logically — would thus be subordinated at once to something general. One can define everything else in the world by ideas, but we must experience our own "I" as something individual within us. Everything that is expressed about the individual in thoughts cannot take up his content into itself; it can only point to it. One says: Look into yourself; there is something for which any concept, any idea, is too poor to encompass in all its incarnate wealth, something that brings forth the ideas out of itself, but that itself has an inexhaustible spring within itself whose content is infinitely more extensive than everything this something brings forth. Stirner's response is: "The individual is a word and with a word one would after all have to be able to think something; a word would after all have to have a thought-content. But the individual is a word without thought; it has no thought-content. But what is its content then if not thought? Its content is one that cannot be there a second time and that consequently can also not be expressed, for if it could be expressed, really and entirely expressed, then it would be there a second time, would be there in the 'expression'... only when nothing of you is spoken out and you are only named, are you recognized as you. As long as something of you is spoken out, you will be recognized only as this something (man, spirit, Christian, etc.)." The individual "I" is therefore that which is everything it is only through itself, which draws the content of its existence out of itself and continuously expands this content from out of itself.

This individual "I" can acknowledge no ethical obligation that it does not lay upon itself. "Whether what I think and do is Christian, what do I care? Whether it is human, liberal, humane, or inhuman, unliberal, inhumane, I don't ask about that. If it only aims at what I want, if I satisfy only myself in it, then call it whatever you like: it's all the same to me ..." "Perhaps, in the very next moment I will turn against my previous thought; I also might very well change my behavior suddenly; but not because it does not correspond to what is Christian, not because it goes against eternal human rights, not because it hits the idea of mankind, humanity, humaneness in the face, but rather — because I am no longer involved, because I no longer enjoy it fully, because I doubt my earlier thought, or I am no longer happy with my recent behavior." The way Stirner speaks about love from this point of view is characteristic. "I also love people, not merely some of them but everyone. But I love them with the consciousness of egoism; I love them because love makes me happy; I love because loving is natural for me, because I like it. I know no 'commandment of love' ..." To this sovereign individual, all state, social, and church organizations are fetters. For, all organizations presuppose that the individual must be like this or like that so that it can fit into the community. But the individual will not let it be determined for him by the community how he should be. He wants to make himself into this or that. J. H. Mackay, in his book Max Stirner, His Life and Work, has expressed what matters to Stirner: "The annihilation, in the first place, of those foreign powers which seek in the most varied ways to suppress and destroy the "I"; and in the second place, the presentation of the relationships of our intercourse with each other, how they result from the conflict and harmony of our interests." The individual cannot fulfill himself in an organized community, but only in free intercourse or association. He acknowledges no societal structure set over the individual as a power. In him everything occurs through the individual. There is nothing fixed within him. What occurs is always to be traced back to the will of the individual. No one and nothing represents a universal will. Stirner does not want society to care for the individual, to protect his rights, to foster his well-being, and so on. When the organization is taken away from people, then their intercourse regulates itself on its own. "I would rather have to rely on people's self-interest than on their 'service of love,' their compassion, their pity, etc. Self-interest demands reciprocity (as you are to me, thus I am to you), does nothing 'for nothing,' and lets itself be won and — bought." Let human intercourse have its full freedom and it will unrestrictedly create that reciprocity which you could set up through a community after all, only in a restricted way. "Neither a natural nor a spiritual tie holds a society (Verein) together, and it is no natural nor spiritual association (Bund). It is not blood nor a belief (i.e., spirit) that brings it about. In a natural association — such as a family, a tribe, a nation; yes, even mankind — individuals have value only as specimens of a species or genus; in a spiritual association — such as a community or church — the individual is significant only as a part of the common spirit; in both cases, what you are as an individual must be suppressed. Only in a society can you assert yourself as an individual, because the society does not possess you, but rather you possess it or use it."

The path by which Stirner arrived at his view of the individual can be designated as a universal critique of all general powers that suppress the "I."

The churches, the political systems (political liberalism, social liberalism, humanistic liberalism), the philosophies — they have all set such general powers over the individual. Political liberalism establishes the "good citizen"; social liberalism establishes the worker who is like all the others in what they own in common; humanistic liberalism establishes the "human being as human being." As he destroys all these powers, Stirner sets up in their ruins the sovereignty of the individual. "What all is not supposed to be my cause! Above all the good cause, then God's cause, the cause of mankind, of truth, of freedom, of humaneness, of justice; furthermore the cause of my folk, of my prince, of my fatherland; finally, of course, the cause of the spirit and a thousand other causes. Only my cause is never supposed to be my cause. — Let us look then at how those people handle their cause for whose cause we are supposed to work, to devote ourselves, and to wax enthusiastic. You know how to proclaim many basic things about God, and for thousands of years have investigated 'the depths of the Divinity' and looked into His heart, so that you are very well able to tell us how God Himself conducts 'the cause of God' that we are called to serve. And you also do not keep the Lord's conduct secret. What is His cause then? Has He, as is expected of us, made a foreign cause, the cause of truth and love, into His own? Such lack of understanding enrages you and you teach us that God's cause is, to be sure, the cause of truth and love, but that this cause cannot be called foreign to Him because God is Himself, in fact, truth and love; you are enraged by the assumption that God could be like us poor worms in promoting a foreign cause as His own. 'God is supposed to take on the cause of truth when He is not Himself the truth?' He takes care only of His cause, but because He is the all in all, everything is also His cause; we, however, we are not the all in all, and our cause is small and contemptible indeed; therefore we must 'serve a higher cause.' — Now, it is clear that God concerns Himself only with what is His, occupies Himself only with Himself, thinks only about Himself, and has His eye on Himself; woe to anything that is not well pleasing to Him. He serves nothing higher and satisfies only Himself. His cause is a purely egoistical cause. How do matters stand with mankind, whose cause we are supposed to make into our own? Is its cause perhaps that of another, and does mankind serve a higher cause? No, mankind looks only at itself, mankind wants to help only mankind, mankind is itself its cause. In order to develop itself, mankind lets peoples and individuals torment themselves in its service, and when they have accomplished what mankind needs, then, out of gratitude, they are thrown by it onto the manure pile of history. Is the cause of mankind not a purely egoistical cause?" Out of this kind of a critique of everything that man is supposed to make into his cause, there results for Stirner that "God and mankind have founded their cause on nothing but themselves. I will then likewise found my cause upon myself, I, who like God am nothing from anything else, I, who am my all, I who am the single one."

That is Stirner's path. One can also take another path to arrive at the nature of the "I." One can observe the "I" in its cognitive activity. Direct your gaze upon a process of knowledge. Through a thinking contemplation of processes, the "I" seeks to become conscious of what actually underlies these processes. What does one want to achieve by this thinking contemplation? To answer this question we must observe: What would we possess of these processes without this contemplation, and what do we obtain through this contemplation? I must limit myself here to a meager sketch of these fundamental questions about world views, and can point only to the broader expositions in my books Truth and Science (Wahrheit und Wissenschaft) and The Philosophy of Spiritual Activity (Die Philosophic der Freiheit).

Look at any process you please. I throw a stone in a horizontal direction. It moves in a curved line and falls to earth after a time. I see the stone at successive moments in different places, after it has first cost me a certain amount of effort to throw it. Through my thinking contemplation I gain the following. During its motion the stone is under the influence of several factors. If it were only under the influence of the propulsion I gave it in throwing it, it would go on forever, in a straight line, in fact, without changing its velocity. But now the earth exerts an influence upon it which one calls gravity. If, without propelling it away from me, I had simply let go of it, it would have fallen straight to the ground, and in doing so its velocity would have increased continuously. Out of the reciprocal workings of these two influences there arises what actually happens. Those are all thought-considerations that I bring to what would offer itself to me without any thinking contemplation.

In this way we have in every cognitive process an element that would present itself to us even without any thinking contemplation, and another element that we can gain only through such thinking contemplation.

When we have then gained both elements, it is clear to us that they belong together. A process runs its course in accordance with the laws that I gain about it through my thinking. The fact that for me the two elements are separated and are joined together by my cognition is my affair. The process does not bother about this separation and joining. From this it follows, however, that the activity of knowing is altogether my affair. Something that I bring about solely for my own sake.

Yet another factor enters in here now. The things and processes would never, out of themselves, give me what I gain about them through my thinking contemplation. Out of themselves they give me, in fact, what I possess without that contemplation. It has already been stated in this essay that I take out of myself what I see in the things as their deepest being. The thoughts I make for myself about the things, these I produce out of my own inner being. They nevertheless belong to the things, as has been shown. The essential being of the things does not therefore come to me from them, but rather from me. My content is their essential being. I would never come to ask about the essential being of the things at all if I did not find present within me something I designate as this essential being of the things, designate as what belongs to them, but designate as what they do not give me out of themselves, but rather what I can take only out of myself.

Within the cognitive process I receive the essential being of the things from out of myself. I therefore have the essential being of the world within myself. Consequently I also have my own essential being within myself. With other things two factors appear to me: a process without its essential being and the essential being through me. With myself, process and essential being are identical. I draw forth the essential being of all the rest of the world out of myself, and I also draw forth my own essential being from myself.

Now my action is a part of the general world happening. It therefore has its essential being as much within me as all other happenings. To seek the laws of human action means, therefore, to draw them forth out of the content of the "I." Just as the believer in God traces the laws of his actions back to the will of his God, so the person who has attained the insight that the essential being of all things lies within the "I" can also find the laws of his action only within the "I." If the "I" has really penetrated into the essential nature of its action, it then feels itself to be the ruler of this action. As long as we believe in a world-being foreign to us, the laws of our action also stand over against us as foreign. They rule us; what we accomplish stands under the compulsion they exercise over us. If they are transformed from such foreign beings into our "I's" primally own doing, then this compulsion ceases. That which compels has become our own being. The lawfulness no longer rules over us, but rather rules within us over the happenings that issue from our "I." To bring about a process by virtue of a lawfulness standing outside the doer is an act of inner unfreedom; to do so out of the doer himself is an act of inner freedom. To give oneself the laws of one's actions out of oneself means to act as a free individual. The consideration of the cognitive process shows the human being that he can find the laws of his action only within himself.

To comprehend the "I" in thinking means to create the basis for founding everything that comes from the "I" also upon the "I" alone. The "I" that understands itself can make itself dependent upon nothing other than itself. And it can be answerable to no one but itself. After these expositions it seems almost superfluous to say that with this "I" only the incarnate real "I" of the individual person is meant and not any general "I" abstracted from it. For any such general "I" can indeed be gained from the real "I" only by abstraction. It is thus dependent upon the real individual. (Benj. R. Tucker and J. H. Mackay also advocate the same direction in thought and view of life out of which my two above-mentioned books have arisen. See Tucker's Instead of a Book and Mackay's The Anarchists.

In the eighteenth century and in the greater part of the nineteenth, man's thinking made every effort to win for the "I" its place in the universe. Two thinkers who are already keeping aloof from this direction are Arthur Schopenhauer and Eduard von Hartmann, who is still vigorously working among us. Neither any longer transfers the full being of our "I," which we find present in our consciousness, as primal being into the outer world. Schopenhauer regarded one part of this "I," the will, as the essential being of the world, and Hartmann sees the unconscious to be this being. Common to both of them is this striving to subordinate the "I" to their assumed general world-being. On the other hand, as the last of the strict individualists, Friedrich Nietzsche, taking his start from Schopenhauer, did arrive at views that definitely lead to the path of absolute appreciation of the individual "I." In his opinion, genuine culture consists in fostering the individual in such a way that he has the strength out of himself to develop everything lying within him. Up until now it was only an accident if an individual was able to develop himself fully out of himself. "This more valuable type has already been there often enough: but as a happy chance, as an exception, never as willed. Rather he was precisely the one feared the most; formerly he was almost the fearful thing; — and out of fear, the opposite type was willed, bred, attained: the domestic animal, the herd animal, the sick animal man, the Christian ..." Nietzsche transfigured poetically, as his ideal, his type of man in his Zarathustra. He calls him the Superman (Übermensch). He is man freed from all norms, who no longer wants to be the mere image of God, a being in whom God is well pleased, a good citizen, and so on, but rather who wants to be himself and nothing more — the pure and absolute egoist.

Reordering of Society
Capital and Credit

From various points of view the opinion has been expressed that all questions of money are so complicated as to be well-nigh impossible to grasp in clear and transparent thoughts. A similar view can be maintained regarding many questions of modern social life. But we should consider the consequences that must follow if men allow their social dealings to be guided by indefinite thoughts; for such thoughts do not merely signify a confusion in theoretic knowledge, they are potent forces in life; their vague character lives on in the institutions that arise under their influence, which in turn result in social conditions making life impossible ...

If we try to go the root of the social question, we are bound to see that even the most material demands can be grappled with only by proceeding to the thoughts that underlie the co-operation of men and women in a community. For example, people closely connected with the land have indicated how, under the influence of modern economic forces, the buying and selling of land has made land into a commodity, and they are of the opinion that this is harmful to society. Yet opinions such as these do not lead to practical results, for men in other spheres of life do not admit that they are justified ... We must take into account how the purely capitalistic tendency affects the valuation of land. Capital creates the laws of its own increase, which in certain spheres no longer accord with an increase on sound lines. This is specially evident in the case of land. Certain conditions may well make it necessary for a district to be fruitful in a particular way-they may be founded on spiritual and cultural peculiarities. But their fulfilment might result in a smaller interest on capital than investment elsewhere. As a consequence of the purely capitalistic tendency the land will then be exploited, not according to these spiritual or cultural points of view, but in such a way that the resulting interest on capital may equal that in other undertakings. And in this way values that may be very necessary to a real civilization are left undeveloped.

It is easy to jump to the conclusion: The capitalistic orientation of economic life has these results, and must therefore be abandoned ... But one who recognizes how modern life works through division of labour and of social function will rather have to consider how to exclude from social life the disadvantages which arise as a by-product of this capitalistic tendency ... The ideal is to work for a structure of society whereby the criterion of increase in capital will no longer be the only power to which production is subject-it should rather be the symptom, which shows that the economic life, by taking into account all the requirements of man's bodily and spiritual nature, is rightly formed and ordered ...

Now it is just in so far as they can be bought and sold for sums of capital in which their specific nature finds no expression, that economic values become commodities. But the commodity nature is only suited to those goods or values which are directly consumed by man. For the valuation of these, man has an immediate standard in his bodily and spiritual needs. There is no such standard in the case of land, nor in the case of means of production. The valuation of these depends on many factors, which only become apparent when one takes into account the social structure as a whole ...

Where 'supply and demand' are the determining factors, there the egoistic type of value is the only one that can come into reckon ing. The 'market' relationship must be superseded by associations regulating the exchange and production of goods by an intelligent observation of human needs. Such associations can replace mere supply and demand by contracts and negotiations between groups of producers and consumers, and between different groups of producers ...

Work done in confidence of the return achievements of others constitutes the giving of *credit* in social life. As there was once a transition from barter to the money system, so there has recently been a progressive transformation to a basis of credit. Life makes it necessary today for one man to work with means entrusted to him by another, or by a community, having confidence in his power to achieve a result. But under the capitalistic method the credit system involves a complete loss of the real and satisfying human relationship of a man to the conditions of his life and work. Credit is given when there is prospect of an increase of capital that seems to justify it; and work is always done subject to the view that the confidence or credit received will have to appear justified in the capitalistic sense. And what is the result? Human beings are subjected to the power of dealings in capital which take place in a sphere of finance remote from life. And the moment they become fully conscious of this fact, they feel it to be unworthy of their humanity ...

A healthy system of giving credit presupposes a social structure which enables economic values to be estimated by their relation to the satisfaction of men's bodily and spiritual needs. Men's economic dealings will take their form from this. Production will be considered from the point of view of needs, no longer by an abstract scale of capital and wages.

Economic life in a threefold society is built up by the cooperation of *associations* arising out of the needs of producers and the interests of consumers. In their mutual dealings, impulses from the spiritual sphere and sphere of rights will play a decisive part. These associations will not be bound to a purely capitalistic standpoint, for one association will be in direct mutual dealings with another, and thus the one-sided interests of one branch of production will be regulated and balanced by those of the other. The responsibility for the giving and taking of credit will thus devolve to the associations. This will not impair the scope and activity of individuals with special faculties; on the contrary, only this method will give individual faculties full scope: the individual is responsible to his association for achieving the best possible results. The association is responsible to other associations for using these individual achievements to good purpose. The individual's desire for gain will no longer be imposing production on the life of the community; production will be regulated by the needs of the community ...

All kinds of dealings are possible between the new associations and old forms of business — there is no question of the old having to be destroyed and replaced by the new. The new simply takes its place and will have to justify itself and prove its inherent power, while the old will dwindle away ... The essential thing is that the threefold idea will stimulate a real social intelligence in the men and women of the community. The individual will in a very definite sense be contributing to the achievements of the whole community ... The individual faculties of men, working in harmony with the human relationships founded in the sphere of rights, and with the production, circulation and consumption that are regulated by the economic associations, will result in the greatest possible efficiency. Increase of capital, and a proper adjustment of work and return for work, will appear as a final consequence ...Whether a man rejects this idea or makes it his own will depend on his summoning the will and energy to work his way through into the sphere of causes. If he does so, he will cease considering external institutions alone; his attention will be guided to the human beings who make the institutions. Division of labour separates men; the forces that come from the three spheres of social life, once these are made independent, will draw them together again ... This inevitable demand of the time is shown in a vivid light by such concrete facts as the continued intensification of the credit system ... In the long run, credit cannot work healthily unless the giver of credit feels himself responsible for all that is brought about through his giving credit. The receiver of credit, through the associations, must give him grounds to justify his taking this responsibility. For a healthy national economy, it is not merely important that credit should further the spirit of enterprise as such, but that the right methods and institutions should exist to enable the spirit of enterprise to work in a socially useful way.

The social thoughts that start from the threefold idea do not aim to replace free business dealings governed by supply and demand by a system of rations and regulations. Their aim is to realize the true relative values of commodities, with the underlying idea that the product of one man's labour should be equivalent in value to all the other commodities that he needs for his consumption during the time he spends in producing it.

Under the capitalistic system, demand may determine whether someone will undertake the production of a certain commodity. But demand alone can never determine whether it will be possible to produce it at a price corresponding to its value in the sense defined above. This can only be determined through methods and institutions by which society in all its aspects will bring about a sensible valuation of the different commodities. Anyone who doubts that this is worth striving for is lacking in vision. For he does not see that, under the mere rule of supply and demand, human needs whose satisfaction would uplift the civilized life of the community are being starved. And he has no feeling for the necessity of trying to include the satisfaction of such needs among the practical incentives of an organised community. The essential aim of the threefold society is to create a just balance between human needs and the value of the products of human work.

Reordering of Society Requirements of Spiritual, Social and Economic Life

In the social movement of the present day there is a great deal of talk about social organization but very little about social and unsocial human beings. Little regard is paid to that 'social question' which arises when one considers that the arrangements of society take their social or antisocial stamp from the people who work in them. Socialist thinkers expect to see in the control of the means of production by the community what will satisfy the needs of the wider population. They take for granted that under such control the co-operation between people must take a social form. They have seen that the industrial system of private capitalism has led to unsocial conditions. They think that if this industrial system were to disappear, the antisocial effects must also end.

Undoubtedly along with the modern capitalistic form of economy there have arisen social ills to the widest extent; but is this any proof that they are a necessary consequence of this economic system? An industrial system can of its own nature do nothing but put men into situations in life that enable them to produce goods for themselves or for others in a useful or a useless manner. The modern industrial system has brought the means of production into the power of individuals or groups of persons. The technical achievements could best be exploited by a concentration of economic power. So long as this power is employed only in the production of goods, its social effect is essentially different from when it trespasses on the fields of civil rights or spiritual culture. And it is this trespassing which in the course of the last few centuries has led to those social ills for whose abolition the modern social movement is pressing. He who is in possession of the means of production acquires economic domination over others. This has resulted in his allying himself with the forces helpful to him in administration and parliaments, through which he was able to procure positions of social advantage over those who were economically dependent on him; and which even in a democratic state bear in practice the character of rights. Similarly this economic domination has led to a monopolizing of the life of spiritual culture by those who held economic power.

Now the simplest thing seems to be to get rid of this economic predominance of individuals, and thereby to do away with their predominance in rights and spiritual culture as well. One arrives at this 'simplicity' of social conception when one fails to remember that the combination of technical and economic activity which modern life demands necessitates allowing the most fruitful expansion possible to individual initiative and personal worth within the business of economic life. The form which production must take under modern conditions makes this a necessity. The individual cannot make his abilities effective in business, if he is tied down in his work and decisions to the will of the community. However dazzling the thought of the individual producing not for himself but for society collectively, yet its justice within certain bounds should not hinder one from also recognizing the other truth, that society collectively is incapable of originating economic decisions that permit of being realized through individuals in the desirable way. Really practical thought, therefore, will not look to find the cure for social ills in a reshaping of economic life that would substitute communal for private management of the means of production. The endeavour should rather be to forestall the ills that can arise through management by individual initiative and personal worth, without impairing this management itself. This is only possible if the relations of civil rights amongst those engaged in industry are not influenced by the interests of economic life, and if that which should be done for people through the spiritual life is also independent of these interests.

Genuine interests of right can only spring up on a ground where the life of rights is separately cultivated, and where the only consideration will be what the rights of a matter are. When people proceed from such considerations to frame rules of right, the rules thus made will take effect in economic life. Then it will not be necessary to place a restriction on the individual acquiring economic power; for such power will only result in his rendering economic achievements proportionate to his abilities, but not in using this to obtain privileged rights ... Only when rights are ordered in a field where a business consideration cannot in any way come into question, where business can procure no power over this system of rights, will the two be able to work together in such a way that men's sense of right will not be injured, nor economic ability be turned from a blessing to a curse for the community as a whole.

When those who are economically powerful are in a position to use their power to wrest privileged rights for themselves, then among the economically weak there will grow up a corresponding opposition to these privileges; and this opposition must as soon as it has grown strong enough lead to revolutionary disturbances. If the existence of a special province of rights makes it impossible for such privileged rights to arise, then disturbances of this sort cannot occur. . . One will never really touch what is working up through the social movement to the surface of modern life, until one brings about social conditions in which, alongside the claims and interests of the economic life, those of rights can find realization and satisfaction on their own independent basis.

In a similar manner must one approach the question of the cultural life, and its connections with the life of civil rights and of industrial economy. The course of the last few centuries has been such that the cultural life itself has been cultivated under conditions which only allowed of its exercising to a limited extent an independent influence upon political life — that of civil rights — or upon economics. One of the most important branches of spiritual culture, the whole manner of education, was shaped by the interests of the civil power. The human being was taught and trained according as state interests required; and state power was reinforced by economic power. If anyone was to develop his capacities within the existing provisions for education, he had to do so on the basis of such finances as his place in life provided. Those spiritual forces that could find scope within the life of political rights or of industry accordingly acquired the stamp of the latter. Any free spiritual life had to forego all idea of carrying its results into the sphere of the state, and could only do so in the economic sphere in so far as this remained outside the sphere of activities of the state. In industry, after all, the necessity is obvious for allowing the competent person to find scope, since all fruitful activity dies out if left solely under the control of the incompetent whom circumstances may have endowed with economic power. If the tendency common amongst socialist thinkers were carried out and economic life were administered after the fashion of the political and legal, then the culture of the free spiritual life would be forced to withdraw altogether from the public field.

But a spiritual life that has to develop apart from civil and industrial realities loses touch with life. It is forced to draw its content from sources that are not in live connection with these realities; and in course of time it works this substance up into a shape which runs on like a sort of animated abstraction along side the actual realities, without having any practical effect on them. And so two different currents arise in spiritual life ... Consider what conceptions of the mind, what religious ideals, what artistic interests form the inner life of the shopkeeper, the manufacturer, or the government official, apart from his daily practical life; and then consider what ideas are contained in those activities expressed in his bookkeeping, or for which he is trained by the education and instruction that prepares him for his profession. A gulf lies between the two currents of spiritual life. The gulf has grown all the wider in recent years because the mode of conception which in natural science is quite justified has become the standard of man's relation to reality. This mode of conception proceeds from the knowledge of laws in things and processes lying outside the field of human activity and influence, so that man is as it were a mere spectator of that which he grasps in the laws of nature ...

A spiritual conception that penetrates to the being of man finds there motives for action which ethically are directly good; for the impulse to evil arises in man only because in his thoughts and sensations he silences the depths of his own nature. Hence social ideas arrived at through the spiritual conception here meant must by their very nature be ethical ideas as well. And not being drawn from thought alone but experienced in life, they have the strength to lay hold on the will and live on in action. For true spiritual conception, social thought and ethical thought flow into one ...

This kind of spirit can, however, thrive only when its growth is completely independent of all authority except such as is derived directly from the spiritual life itself. Legal regulations by the civil state for the nurture of the spirit sap the strength of the forces of spiritual life, whereas a spiritual life left to its own inherent interests and impulses will reach out into everything that man performs in social life ...

If the life of the spirit be a free one, evolved only from impulses within itself, then civil life will thrive in proportion as people are educated intelligently from living spiritual experience in the adjustment of their relationships of rights; and economic life will be fruitful in the measure in which men's spiritual nature has developed their capacities for it ...

Because the spirit at work in civil life and the round of industry is no longer one through which the spiritual life of the individual finds a channel, he sees himself in a social order which gives him, as individual, no scope civically nor economically. People who do not see this clearly will always object to a view of the social organism divided into three independently functioning systems of the cultural life, the rights state and the industrial economy, that such a differentiation would destroy the necessary unity of communal life. One must reply to them that this unity is destroying itself, in the effort to maintain itself intact ... It is just in separation that they will turn to unity, whereas in an artificial unity they become estranged.

Many socialist thinkers will dismiss such an idea with the phrase that conditions of life worth striving for cannot be brought about by this organic membering of society, but only through a suitable economic organization. They overlook the fact that the men at work in their organization are endowed with wills. If one tells them so they will smile, for they regard it as self-evident. Yet they envisage a social structure in which this 'self-evident' fact is left out of account. Their economic organization is to be controlled by a communal will, which must be the resultant wills of the people in the organization. These individual wills can never find scope, if the communal will is derived entirely from the idea of economic organization ...

Most people today still lack faith in the possibility of establishing a socially satisfying order of society based on individual wills, because such a faith cannot come from a spiritual life dependent on the life of the state and of the economy. The kind of spirit that develops not in freedom out of the life of the spirit itself but out of an external organization simply does not know what the potentialities of the spirit are. It looks round for something to direct it, not knowing how the spirit directs itself if only it can draw its strength from its own resources. For the new shaping of the social order, goodwill is not the only thing needed. It needs also that courage which can be a match for the lack of faith in the spirit's power. A true spiritual conception can inspire this courage; for it feels able to bring forth ideas that not only serve to give the soul its inward orientation, but which in their very birth bring with them seeds of life's practical configuration. The will to go down into the deep places of the spirit can become a will so strong as to bear a part in every thing that man performs ...

The experiments now being made to solve the social question afford such unsatisfactory results because many people have not yet become able to see what the true gist of the problem is. They see it arise in economic regions, and look to economic institutions to provide the answer. They think they will find the solution in economic transformations. They fail to recognize that these transformations can only come about through forces released from within human nature itself in the uprising of a new spiritual life and life of rights in their own independent realms.

Reordering of Society
The Fundamental Social Law

Briefly as the subject must be dealt with, there will always be some people whose feeling will lead them to recognize the truth of what it is impossible to discuss in all its fullness here. There is a fundamental social law which spiritual science teaches, and which is as follows:

'The well-being of a community of people working together will be the greater, the less the individual claims for himself the proceeds of his work, i.e. the more of these proceeds he makes over to his fellow-workers, the more his own needs are satisfied, not out of his own work but out of the work done by others'.

Every arrangement in a community that is contrary to this law will inevitably engender somewhere after a while distress and want. It is a fundamental law, which holds good for all social life with the same absoluteness and necessity as any law of nature within a particular field of natural causation. It must not be supposed, however, that it is sufficient to acknowledge this law as one for general moral conduct, or to try to interpret it into the sentiment that everyone should work in the service of his fellow men. No, this law only lives in reality as it should when a community of people succeeds in creating arrangements such that no one can ever claim the fruits of his own labour for himself, but that these go wholly to the benefit of the community. And he must himself be supported in return by the labours of his fellow men. The important point is, therefore, that working for one's fellow men and obtaining so much income must be kept apart, as two separate things.

Self-styled 'practical people' will of course have nothing but a smile for such 'outrageous idealism'. And yet this law is more practical than any that was ever devised or enacted by the 'practicians'. Anyone who really examines practical life will find that every community that exists or has ever existed anywhere has two sorts of arrangements, of which the one is in accordance with this law and the other contrary to it. It is bound to be so everywhere, whether men will it or not. Every community would indeed fall to pieces at once, if the work of the individual did not pass over into the totality. But human egoism has from of old run counter to this law, and sought to extract as much as possible for the individual out of his own work. And what has come about from of old in this way due to egoism has alone brought want, poverty and distress in its wake. This simply means that the part of human arrangements brought about by 'practicians' who calculated on the basis of either their own egotism or that of others must always prove impractical.

Now naturally it is not simply a matter of recognizing a law of this kind, but the real practical part begins with the question: How is one to translate this law into actual fact? Obviously this law says nothing less than this: man's welfare is the greater, in proportion as egoism is less. So for its translation into reality one must have people who can find their way out of egoism. In practice, however, this is quite impossible if the individual's share of weal and woe is measured according to his labour. He who labours for himself must gradually fall a victim to egoism. Only one who labours solely for the rest can gradually grow to be a worker without egoism.

But there is one thing needed to begin with. If any man works for another, he must find in this other man the reason for his work; and if anyone is to work for the community, he must perceive and feel the value, the nature and importance, of this community. He can only do this when the community is something quite different from a more or less indefinite summation of individual men. It must be informed by an actual spirit, in which each single one has his part. It must be such that each one says: 'It is as it should be, and I will that it be so'. The community must have a spiritual mission, and each individual must have the will to contribute towards the fulfilling of this mission. All the vague abstract ideals of which people usually talk cannot present such a mission. If there be nothing but these, then one individual here or one group there will be working without any clear overview of what use there is in their work, except it being to the advantage of their families, or of those particular interests to which they happen to be attached. In every single member, down to the most solitary, this spirit of the community must be alive ...

No one need try to discover a solution of the social question that shall hold good for all time, but simply to find the right form for his social thoughts and actions in the light of the immediate need of the time in which he lives. Indeed there is today no theoretical scheme which could be devised or carried into effect by any one person which in itself could solve the social question. For this he would need to possess the power to force a number of people into the conditions which he had created. But in the present day any such compulsion is out of the question. The possibility must be found of each person doing of his own free will that which he is called upon to do according to his strength and abilities. For this reason there can be no possible question of ever trying to work on people theoretically, by merely indoctrinating them with a view as to how economic conditions might best be arranged. A bald economic theory can never act as a force to counteract the powers of egoism. For a while such an economic theory may sweep the masses along with a kind of impetus that appears to resemble idealism; but in the long run it helps nobody. Anyone who implants such a theory into a mass of people without giving them some real spiritual substance along with it is sinning against the real meaning of human evolution. The only thing which can help is a spiritual world-conception which of itself, through what it has to offer, can live in the thoughts, in the feelings, in the will — in short, in a man's whole soul ...

The recognition of these principles means, it is true, the loss of many an illusion for various people whose ambition it is to be popular benefactors. It makes working for the welfare of society a really difficult matter — one of which the results, too, may in certain circumstances comprise only quite tiny part-results. Most of what is given out today by whole parties as panaceas for social life loses its value, and is seen to be a mere bubble and hollow phrase, lacking in due knowledge of human life. No parliament, no democracy, no popular agitation can have any meaning for a person who looks at all deeper, if they violate the law stated above; whereas everything of this kind may work for good if it works on the lines of this law. It is a mischievous delusion to believe that particular persons sent up to some parliament as delegates from the people can do anything for the good of mankind, unless their activity is in conformity with the fundamental social law.

Wherever this law finds outer expression, wherever anyone is at work on its lines — so far as is possible in that position in which he is placed within the community — good results will be attained, though it be but in the single case and in never so small a measure. And it is only a number of individual results attained in this way that will together combine to the healthy collective progress of society.

The healthy social life is found   When in the mirror of each human soul The whole community is shaped,   And when in the community   Lives the strength of each human soul.

The Human Soul
in the Twilight of Dreams

If, within the limits of ordinary consciousness, the human being wishes to study his soul, it will not suffice for him simply to direct his mind's eye backward, so to speak, in order to discern by introspection his nature as someone who looks out upon the world. He will see nothing new by this means. He will perceive himself in his capacity as a spectator of the world — merely from a different direction. In his waking life man is almost entirely occupied with the external world. He lives by his senses. In their impressions the external world continues to live in his inner life. Thoughts weave into these impressions. The outer world lives in the thoughts as well. Only the force with which the world is grasped in thoughts can be experienced as man's autonomous being. The sensation of this force, however, is of an entirely general and vague character. By means of ordinary consciousness one can differentiate nothing within this sensation. If one had to discern the human soul in it, one would obtain no more about the soul than a vague sensation of self; one would be unable to identify what it was.

What is unsatisfying about self-observation along these lines is that the nature of the soul promptly eludes the attempt to grasp it. Because of this drawback, people who seriously strive after knowledge may be driven to despair of it entirely.

Thoughtful people, therefore, have almost always sought knowledge of the human soul in ways other than such selfobservation. In the realm of sense perception and ordinary thinking they have felt that the vague sensation of self is surrendered to the body. They have realized that the soul, so long as it remains in this surrender, can learn nothing of its own nature through self-observation.

A realm to which this feeling points is that of the dream. People have become aware that the world of images the dream conjures up has some connection with the vague sensation of self. This appears, as it were, as an empty canvas on which the dream paints its own pictures. And then it is realized that the canvas is really itself the painter painting on and within itself.

Dreaming thus becomes for them the fleeting activity of man's inner life, which fills the soul's vague feeling of self with content. A questionable content — but the only one to be had at the outset. It is a view, lifted out of the brightness of ordinary consciousness and thrust into the twilight of semi-consciousness. Yet this is the only form attainable in everyday life.

Despite this dimness, however, there occurs — not in thinking self-observation to be sure, but in an inward touching of the self— something very significant. A kinship between dreaming and creative fantasy can inwardly be touched [seelisch ertasten]. One has the feeling that the airy pictures of a dream are the same as those of creative fantasy, though the latter are controlled by the body from within. And this inner body [Korper-Innere] compels the dream-picturing power to desist from its arbitrariness and to transform itself into an activity that emulates, albeit in a free manner, what exists in the world of the senses.

Once one has struggled through to such a touching of the inner world, one soon advances a step further. One becomes aware how the dream-picturing power can form a still closer connection with the body. One sees this foreshadowed in the activity of recollection, of memory. In memory the body compels the dream-picturing power to an even stronger fidelity to the outer world than it does in fantasy.

If this is understood, then there remains but one step to the recognition that the dream-picturing force of the soul also lies at the basis of ordinary thinking and sense perception. It is then entirely surrendered to the body, while in fantasy and memory it still reserves something of its own weaving.

This, then, justifies the assumption that in dreaming the soul frees itself from the state of bondage to the body and lives according to its own nature.

Thus the dream has become the field of inquiry for many searchers after the soul.

It relegates man, however, to a quite uncertain province. In surrendering to the body, the human soul becomes harnessed to the laws which govern nature. The body is a part of nature. Insofar as the soul surrenders to the body, it binds itself at the same time to the regularity of nature. The means whereby the soul adapts to the existence of nature is experienced as logic. In logical thinking about nature the soul feels secure. But in the power of dream-making it tears itself away from this logical thinking about nature. It returns to its own sphere. Thereby it abandons, as it were, the welltended and well-trodden pathways of the inner life and sets forth on the flowing, pathless sea of spiritual existence.

The threshold of the spiritual world seems to have been crossed; after the crossing, however, only the bottomless, directionless spiritual element presents itself. Those who seek to cross the threshold in this way find the exciting but also doubt-riddled domain of the soul life.

It is full of riddles. At one time it weaves the external events of life into airy connections that scorn the regularity of nature; at another it shapes symbols of inner bodily processes and organs. A too violently beating heart appears in the dream as an oven; aching teeth as a fence with pickets in disrepair. What is more, man comes to know himself in a peculiar way. His instinctive life takes shape in the dream in images of reprehensible actions which, in the waking state, he would strongly resist. Those dreams that have a prophetic character arouse special interest among students of the soul, as do those in which the soul dreams up capacities that are entirely absent in the waking state.

The soul appears released from its bondage to bodily and natural activity. It wants to be independent, and it prepares itself for this independence. As soon as it tries to become active, however, the activity of the body and of nature follow it. The soul will have nothing to do with nature's regularity; but the facts of nature appear in dreams as travesties of nature. The soul is interested in the internal bodily organs or bodily activities. It cannot, however, make clear pictures of these organs or bodily activities, but only symbols which bear the character of arbitrariness. Experience of external nature is torn away from the certainty in which sense perception and thinking place it. The inner life of what is human begins; it begins, however, in dim form. Observation of nature is abandoned; observation of the self is not truly achieved. The investigation of the dream does not place man in a position to view the soul in its true form. It is true this is spiritually more nearly comprehensible through dreams than through thinking self-observation; it is, however, something he should actually see but can only grope after as if through a veil.

The following section will speak about the perception of the soul through spirit knowledge.

The Human Soul
in Courage and Fear

The habits of thinking that have come to be accepted in the modern study of nature [Naturerkenntnis] can yield no satisfying results for the study of the soul. What one would grasp with these habits of thinking must either be spread out in repose before the soul or, if the object of knowledge is in movement, the soul must feel itself extricated from this movement. For to participate in the movement of the object of knowledge means to lose oneself in it, to transform oneself, so to speak, into it.

How should the soul grasp itself, however, in an act of knowing in which it must lose itself? It can expect selfknowledge only in an activity in which, step by step, it comes into possession of itself.

This can only be an activity that is creative. Here, however, a cause for uncertainty arises at once for the knower. He believes he will lapse into personal arbitrariness.

It is precisely this arbitrariness that he gives up in the knowledge of nature. He excludes himself and lets nature hold sway. He seeks certainty in a realm which his individual soul being does not reach. In seeking self-knowledge he cannot conduct himself in this way. He must take himself along wherever he seeks to know. He therefore can find no nature on his path to self-knowledge. For where she would encounter him, there he is no longer to be found.

This, however, provides just the experience that is needed with regard to the spirit. One cannot expect other than to find the spirit when, through one's own activity, nature, as it were, melts away; that is, when one experiences oneself ever more strongly in proportion to one's feeling this melting away.

If one fills the soul with something that afterward proves to be like a dream in its illusory character, and one experiences the illusory in its true nature, then one becomes stronger in one's own experience of self. In confronting a dream, one's thinking corrects the belief one has in the dream's reality while dreaming. Concerning the activity of fantasy, this correction is not needed because one did not have this belief. Concerning the meditative soul activity, to which one devotes oneself for spirit-knowledge, one cannot be satisfied with mere thought correction. One must correct by experiencing. One must

first create the illusory thinking with one's activity and then extinguish it by
a different, equally strong, activity.

In this act of extinguishing, another activity awakens, the spirit-knowing
activity. For if the extinguishing is real, then the force for it must come from
an entirely different direction than from nature. With the experienced
illusion one has dispersed what nature can give; what inwardly arises during
the dispersion is no longer nature.

With this activity something is needed that does not come into
consideration in the study of nature: inner courage. With it one must take
hold of what inwardly arises. In the study of nature one needs to hold nothing
inwardly. One lets oneself be held by what is external. Inner courage is not
needed here. One forgets it. This forgetting then causes anxiety when the
spiritual is to enter the sphere of knowledge. Fear is felt because one might
grope in a void if one no longer could hold onto nature.

This fear meets one at the threshold to spirit knowledge. And fear causes
one to recoil from this knowledge. One now becomes creative in recoiling
instead of in pressing forward. One does not allow the spirit to shape creative
knowledge in oneself; one invents for oneself a sham logic for disputing the
justification of spirit knowledge. Every possible sham reason is brought
forward to spare one from acknowledging the spiritual, because one retreats
trembling in fear of it.

Instead of spirit knowledge, then, there arises out of the creative force that
which now appears in the soul when it draws back from nature, the enemy of
spirit knowledge: first, as doubt concerning all knowledge that extends
beyond nature; and then, as the fear grows, as an anti-logic that would banish
all spirit knowledge to the realm of the fantastic.

Whoever has learned to move cognitively in the spirit often sees in the
refutations of this knowledge its strongest evidence; for it becomes clear to
him how in the soul, step by step, the refuter chokes down his fear of the
spirit, and how in choking it he creates this sham logic. With such a refuter
there is no point in arguing, for the fear befalling him arises in the
subconscious. The consciousness tries to rescue itself from this fear. It feels
at first that should this anxiety arise, it would inundate the whole inner
experience with weakness. It is true, the soul cannot escape from this
weakness, for one feels it rising up from within. If one ran away it would
follow one everywhere. He who proceeds further in the knowledge of nature
and, in his dedication to it feels obliged to preserve his own self, never
escapes from this fear if he cannot acknowledge the spirit. Fear will
accompany him, unless he is willing to give up the knowledge of nature along

with spirit knowledge. He must somehow rid himself of this fear in his pursuit of the science of nature. In reality he cannot do so. The fear is produced in the: subconscious during the study of nature. It continually attempts to rise up out of the subconscious into consciousness. Therefore one refutes in the thought world what one cannot remove from the reality of soul experience.

And this refutation is an illusory layer of thought covering the subconscious fear. The refuter has not found the courage to come to grips with the illusory, just as in the meditative life he has to obliterate illusion in order to attain spiritual reality. For this reason he interposes the false arguments of his refutation into that region of the life of the soul that now arises. They soothe his consciousness; he ceases to feel the fear that, all the same, remains in his subconscious.

The denial of the spiritual world is a desire to run away from one's own soul. This, however, represents an impossibility. One must remain with oneself. And because one may run away but not escape from oneself, one takes care that in running one loses sight of oneself. It is the same with the entire human being in the soul realm, however, as it is with the eye with a cataract. The eye can then no longer see. It is darkened within itself.

So, too, the denier of spirit knowledge darkens his soul. He causes its darkening through sham reasoning born of fear. He avoids healthy clarification of the soul; he creates for himself an unhealthy soul darkening. The denial of spirit knowledge has its origin in a cataract affliction of the soul.

Thus one is ultimately led to the inner spiritual strength of the soul when one is willing to see the justification of spirit knowledge. And the way to such a knowledge can be had only through the strengthening of the soul. The meditative activity, preparing the soul for spirit knowledge, is a gradual conquest of the soul's "fear of the void." This void, however, is only a "void of nature," in which the "fullness of the spirit" can manifest itself if one wishes to take hold of it. Nor does the soul enter this "fullness of the spirit" with the arbitrariness it has when acting through the body in natural life; the soul enters this fullness at the moment when the spirit reveals to the soul the creative will, before which the arbitrariness, existing only in natural life, dissolves in the same way as nature herself dissolves.

The Human Soul
in the Light of Spirit Vision

If one resorts to dream phenomena in order to acquire knowledge of the soul's nature, one ultimately is forced to admit that the object of one's search is wearing a mask. Behind the symbolizations of bodily conditions and processes, behind the fantastically connected memory experiences, one may surmise the soul's activity. It cannot be maintained, however, that one is face to face with the true form of the soul.

On awaking, one realizes how the active part of the dream is interwoven with the function of the body and thereby subject to the external world of nature. Through the backward-directed view of self-observation one sees in the soul life only the images of the external world, not the life of the soul itself. The soul eludes the ordinary consciousness at the very moment one would grasp it cognitively.

By studying dreams one cannot hope to arrive at the reality of the soul element. In order to preserve the soul activity in its innate form one would have to obliterate, through a strong inner activity, the symbolizations of the bodily conditions and processes, along with the memory of past experiences. Then one would have to be able to study that which had been retained. This is impossible. For the dreamer is in a passive state. He cannot undertake any autonomous activity. With the disappearance of the soul's mask, the sensation of one's own self disappears also.

It is different with the waking soul life. There the autonomous activity of the soul can not only be sustained when one erases all one perceives of the external world; it can also be strengthened in itself.

This happens if, while awake in the forming of mental pictures, one makes oneself as independent of the external world of the senses as one is in a dream. One becomes a fully conscious, wakeful imitator of the dream. Thereby, however, the illusory quality of the dream falls away. The dreamer takes his dream pictures for realities. If one is awake one can see through their unreality. No healthy person when awake and imitating the dream will take his dream images for realities. He will remain conscious of the fact that he is living in self-created illusions.

He will not be able to create these illusions, however, if he merely remains at the ordinary level of consciousness. He must see to it that he strengthens this consciousness. He can achieve this by a continually renewed self-kindling of thinking from within. The inner soul activity grows with these repeated kindlings. (I have described in detail the appropriate inner activity in my books Knowledge of the Higher Worlds and its Attainment and An Outline of Occult Science).

In this way the work of the soul during the twilight of dreams can be brought into the clear light of consciousness. One accomplishes thereby the opposite of what happens in suggestion or auto-suggestion. With these, something out of the semi-darkness and within the semi-darkness is shifted into the soul-life, which is then taken to be reality. In the fully circumspect activity of the soul just described, something is placed before one's inner view in clear consciousness, something that one regards, in the fullest sense of the word, only as illusion.

One thus arrives at compelling the dream to manifest itself in the light of consciousness. Ordinarily this occurs only in the diminished half-awake consciousness. It shuns the clarity of consciousness. It disappears in its presence. The strengthened consciousness holds it fast.

In holding the dream fast it does not gain in strength. On the contrary it diminishes in strength. Consciousness, however, is thereby induced to supply its own strength. The same thing happens here in the soul. It is just as it is when, in physical life, one transforms a solid into steam. The solid has its own boundaries on all sides. One can touch these boundaries. They exist in themselves. If one transforms the solid into steam, then one must enclose it within solid boundaries so that it will not escape. Similarly the soul, if it would hold fast the dream while awake, must shape itself, as it were, into a strong container. It must strengthen itself from within.

The soul does not need to effect this strengthening when it perceives the images of the external world. Then the relationship of the body to the external world takes care that the soul is aroused to retain these images. If, however, the waking soul is to dream in sensory unreality, then it must hold fast this sensory unreality by its own strength.

In the fully conscious representation [Vorstellen] of sensory unreality one develops the strength to behold the spiritual reality.

In the dream state the autonomous activity of the soul is weak. The fleeting dream content overpowers this autonomous activity. This supremacy of the dream causes the soul to take the dream for reality. In ordinary waking consciousness the autonomous activity experiences itself as reality along with the reality of the sense world. This autonomous activity, however, cannot behold [anschauen] itself; its vision is occupied with the images of sense reality. If the autonomous activity learns to maintain itself by consciously filling itself with content unreal to the senses, then, little by little, it also brings to life self-contemplation [Anschaung] within itself. Then, it does not simply direct its gaze away from outer observation and back upon itself; it strides as soul activity backward and discovers itself as spiritual entity; this now becomes the content of its vision [Anschaung].

While the soul thus discovers itself within itself, the nature of dreaming is even more illumined for it. The soul discerns clearly what before it could only surmise: that dreaming does not cease in the waking state. It continues. The feeble activity of the dream, however, is drowned by the content of sense perception. Behind the brightness of consciousness, filled with the images of sense reality, there glimmers a dream world. And this world, while the soul is awake, is not illusory like the dream world of semi-consciousness. In the waking state man dreams — beneath the threshold of consciousness — about the inner processes of his body. While the external world is seen through the eye and is present in [vorgestallt] the soul, there lives in the background the dim dream of inner occurrences. Through the strengthening of the autonomous activity of the soul the vision of the external world is gradually dampened to the dimness of dream, and the vision of the inner world, in its reality, brightens.

In its vision of the external world the soul is receptive; it experiences the external world as the creative principle and the soul's own content as created in the image of the external world. In the inner vision, the soul recognizes itself to be the creative principle. And one's own body is revealed as created in the image of the soul. Thoughts of the external world are to be felt [empfunden] as images of the beings and processes of the external world. To the soul's true vision, achieved in the way described, the human body can be felt [empfunden] only as the image of the human soul which is spiritual.

In dreams, the soul activity is loosened from its firm union with the body, which it maintains in the ordinary waking state; it still retains, however, the loose relationship that fills it with the symbolic images of bodily senses and with the memory experiences that also are acquired through the body. In spiritual vision of itself the soul so grows in strength that its own higher reality becomes discernible, and the body becomes recognizable in its character of a reflected image of this reality.

The Human Soul
on the Path to Self-Observation

In a dream the soul comprehends itself in a fleeting form, which is really a mask. In dreamless sleep it apparently loses itself entirely. In spiritual self-contemplation [Anschaung], which is achieved through circumspect reconstruction [besonnen] of the dream-state, the soul comes into its own as a creative being, of which the physical body is the reflected image.

A dream, however, arises out of sleep. Whoever undertakes to raise the dream up into the clear light of consciousness must also feel the incentive to go still further. He does this when he tries consciously to experience dreamless sleep.

That seems to be impossible, precisely because in sleep consciousness ceases. The desire consciously to experience unconsciousness seems like folly.

The folly, however, takes on another light when one confronts the memories one can follow from a given point of time backward to one's last awakening. To do so one must proceed in such a way as to connect the memory pictures vividly with that which they recall. Then, if one tries — working backward — to proceed to the next conscious memory picture before that, this will be found before the last falling asleep. If one has really made the connection vivid with what is recalled, there arises an inner difficulty. One cannot join up the memory picture after awaking with the one before falling asleep.

Ordinary consciousness gets one over this difficulty by not vividly connecting what is recalled, but simply placing the waking image next to the image one has on falling asleep. The person who has raised his consciousness to a high degree of sensitiveness, however, through conscious imitation of the dream, finds that the two images fall apart from one another [fallen ausenander]. For him an abyss lies between them, but because he notices this abyss it already begins to fill itself up. For his self-awareness the dreamless sleep ceases to be an empty passage of time. Out of it there emerges like a memory a spiritual content of the "empty time," like a memory, it is true, of something that ordinary consciousness had not contained before. Even so this memory points to an experience of one's own soul like an ordinary memory. The soul, however, really looks thereby into that which in ordinary experience — in dreamless sleep — occurred unconsciously.

On this path the soul looks still more deeply within than it does in the condition that arises as a result of the conscious dream imitation. In this condition the soul beholds its own body-forming being. Through the conscious penetration of dreamless sleep, the soul perceives itself in its own being, completely detached from the body.

Now, however, the soul beholds not only the forming of the body but also, beyond that, the formation of its own willing [Wollen].

The inner nature of the will remains as unknown to ordinary consciousness as the events of dreamless sleep. One experiences a thought that contains the intention of the will. This thought sinks into the obscure world of the feelings and disappears into the darkness of the bodily processes. It emerges again as the external bodily process of an arm movement that is comprehended anew through a thought. Between the two thought contents there lies something like the sleep between the thoughts before falling asleep and those after waking.

Now as the inner working of the soul upon the body becomes comprehensible to the first level of vision, so does the will over and above the body to the second. The soul can follow the path to behold its inner working upon the body's organic development; and it can take the other path by which it learns to comprehend how the soul works on the body in such a way as to extract the will from it.

And just as dreaming lies between sleeping and waking, so feeling lies between willing and thinking. On the same path that leads to the illumination of the will process lies the illumination of the world of feeling also.

In the first kind of vision the soul's inner working on the organism is revealed. In the second the soul penetrates to the will. But an inner activity must precede the outward manifestation of the will. Before the arm can be raised, the creative current must flow into it so that in its metabolic processes, which run on quietly, processes are inserted that are clearly the result of feeling. Feeling is a willing that remains enclosed within the human being, a willing that is arrested at its inception.

The processes inserted into the body for feeling and willing reveal themselves for the second stage of vision as processes that are in opposition to those that support life. They are destructive processes. In the constructive processes life prospers; but the soul withers in them. The life of the body, which itself is built up by the soul, must be broken down so that the nature and activity of the soul can unfold out of the body.

To spiritual vision the working of the soul on the body is like a memory of something that the soul had first to accomplish before it could exist in its own activity.

Thereby, however, the soul experiences itself as a purely spiritual being that has let the forming of the body take precedence to the soul's own activity in order to have the body become the basis for the soul's inherent, purely spiritual development. The soul first devotes its creative effort to the body so that, after this has been done sufficiently, the soul can manifest itself in free spirituality.

And this development of the soul begins already with thinking that results from the perception of the senses. When one perceives an object, the soul commences its activity. It shapes the corresponding part of the body in such a way that it becomes adapted for developing, in the form of thought, a mirror image of the object. In experiencing this mirror image, the soul beholds the result of its own activity.

One will never find the spiritual nature of the soul by philosophizing about the thoughts that arise before ordinary consciousness. The spiritual activity of the soul does not lie in these thoughts but behind them. It is true that the thoughts which the soul experiences are the result of the brain's activity. The brain's activity, however, is first the result of the spiritual activity of the soul. In misunderstanding this fact lies what is unsound in the materialistic world view. This view is right when it demonstrates from every possible scientific presupposition that thoughts are the result of the brain's activity. Any other view that seeks to contradict this will always run up against the claims of materialism. The activity of the brain, however, is the product of the activity of the spirit. To realize this it is not sufficient to look back into the inner being of man. In doing so one encounters thoughts. And these contain only a pictorial reality. This pictorial reality is the product of the physical body. In observing retrospectively one must bring to life reinforced and strengthened soul capacities. One must wrest the dreaming soul from the twilight of the dream; then it will not evaporate into fantasies, but rather lay its mask aside so as to appear as a being active spiritually in the body. One must wrest the sleeping soul from the darkness of sleep; then the soul does not lose sight of itself but faces itself as an actual spiritual entity, which in the act of willing, by means of the bodily organism, creates above and beyond this body.

Answers to Some Questions
Concerning Karma

The following question has been asked: "According to the law of reincarnation, we are required to think that the human individuality possesses its talents, capacities, and so forth, as an effect of its previous lives. Is this not contradicted by the fact that such talents and capacities, for instance moral courage, musical gifts, and so forth, are directly inherited by the children from their parents?"

Answer: If we rightly conceive of the laws of reincarnation and karma, we cannot find a contradiction in what is stated above. Only those qualities of the human being which belong to his physical and ether body can be directly passed on by heredity. The ether body is the bearer of all life phenomena (the forces of growth and reproduction). Everything connected with this can be directly passed on by heredity. What is bound to the so-called soul-body can be passed on by heredity to a much lesser degree. This constitutes a certain disposition in the sensations. Whether we possess a vivid sense of sight, a well-developed sense of hearing, and so forth, may depend upon whether our ancestors have acquired such faculties and have passed them on to us by heredity. But nobody can pass on to his offsprings what is connected with the actual spiritual being of man, that is, for instance, the acuteness and accuracy of his life of thought, the reliability of his memory, the moral sense, the acquired capacities of knowledge and art.

These are qualities which remain enclosed within his individuality and which appear in his next incarnation as capacities, talents, character, and so forth. — The environment, however, into which the reincarnating human being enters is not accidental, but it is necessarily connected with his karma. Let us assume a human being has acquired in his previous life the capacity for a morally strong character. It is his karma that this capacity should unfold in his next incarnation. This would not be possible if he did not incarnate in a body which possesses a quite definite constitution. This bodily constitution, however, must be inherited from the forebears. The incarnating individuality strives, through a power of attraction inherent in it, toward those parents who are capable of giving it the suitable body. This is caused by the fact that, already before reincarnating, this individuality connects itself with the forces of the astral world which strive toward definite physical conditions. Thus the human being is born into that family which is able to transmit to him by heredity the bodily conditions which correspond to his karmic potentialities. It then looks, if we go back to the example of moral courage, as if the latter itself had been inherited from the parents. The truth is that man, through his individual being, has searched out that family which makes the unfoldment of moral courage possible for him. In addition to this it may be possible that the individualities of the children and the parents have already been connected in previous lives and for that very reason have found one another again. The karmic laws are so complicated that we may never base a judgment upon outer appearances. Only a person to whose spiritual sense-organs the higher worlds are at least partially manifest may attempt to form such a judgment. Whoever is able to observe the soul organism and the spirit, in addition to the physical body, is in a position to discriminate between what has been passed on to the human being by his forebears and what is his own possession, acquired in previous lives. For ordinary vision these things are not clearly distinguishable, and it may easily appear as if something were merely inherited which in reality is karmicly determined. — It is a thoroughly wise expression which states that children are "given" to their parents. In respect of the spirit this is absolutely the case. And children with certain spiritual qualities are given to them for the very reason that they, the parents, are capable of giving the children the opportunity to unfold these spiritual qualities.

Question: "Does Anthroposophy attribute no significance to 'chance'? I cannot imagine that it can be predestined by the karma of each individual person when five hundred persons are killed at the same time in a theater fire."

Answer: The laws of karma are so complicated that we should not be surprised when to the human intellect some fact appears at first as being contradictory to the general validity of this law. We must realize that this intellect is schooled by our physical world, and that, in general, it is accustomed to admit only what it has learned in this world. The laws of karma, however, belong to higher worlds. Therefore, if we try to understand an event which meets the human being as being brought about by karma in the same way in which justice is applied in the purely earthly-physical life, then we must of necessity run up against contradictions. We must realize that a common experience which several people undergo in the physical world may, in the higher world, mean something completely different for each individual person among them. Naturally, the opposite may also be true:

common interrelations may become effective in common earthly experiences. Only one gifted with clear vision in the higher worlds can give information about particular cases. If the karmic interrelations of five hundred people become effective in the common death of these people in a theater fire, the following instances may be possible:

First: Not a single one of the five hundred people need be karmicly linked to the other victims. The common disaster is related in the same way to the karmas of each single person as the shadow-image of fifty people on a wall is related to the worlds of thought and feeling of these persons. These people had nothing in common an hour ago; nor will they have anything in common an hour hence. What they experienced when they met at the same place will have a special effect for each one of them. Their association is expressed in the above-mentioned common shadow-image. Whoever were to attempt to conclude from this shadow-image that a common bond united these people would be decidedly in error.

Second: It is possible that the common experience of the five hundred people has nothing whatsoever to do with their karmic past, but that, just through this common experience, something is prepared which will unite them karmicly in the future. Perhaps these five hundred people will, in future ages, carry out a common undertaking, and through the disaster have been united for the sake of higher worlds. The experienced spiritual-scientist is thoroughly acquainted with the fact that many societies, formed today, owe their origin to the circumstance of a common disaster experienced in a more distant past by the people who join together today.

Third: The case in question may actually be the effect of former common guilt of the persons concerned. There are, however, still countless other possibilities. For instance, a combination of all three possibilities described might occur.

It is not unjustifiable to speak of "chance" in the physical world. And however true it is to say: there is no "chance" if we take into consideration all the worlds, yet it would be unjustifiable to eradicate the word "chance" if we are merely speaking of the interlinking of things in the physical world. Chance in the physical world is brought about through the fact that things take place in this world within sensible space. They must, in as far as they occur within this space, also obey the laws of this space. Within this space, things may outwardly meet which have inwardly nothing to do with each other. The causes which let a brick fall from a roof, injuring me as I pass by, do not necessarily have anything to do with my karma which stems from my past. Many people commit here the error of imagining karmic relations in too simple a fashion. They presume, for instance, that if a brick has injured a person, he must have deserved this injury karmicly. But this is not necessarily so. In the life of every human being events constantly take place which have nothing at all to do with his merits or his guilt in the past. Such events find their karmic adjustment in the future. If something happens to me today without being my fault, I shall be compensated for it in the future. One thing is certain: nothing remains without karmic adjustment. However, whether an experience of the human being is the effect of his karmic past or the cause of his karmic future will have to be determined in every individual instance. And this cannot be decided by the intellect accustomed to dealing with the physical world, but solely by occult experience and observation.

Question: "Is it possible to understand, according to the law of reincarnation and karma, how a highly developed human soul can be reborn in a helpless, undeveloped child? To many a person the thought that we have to begin over and over again at the childhood stage is unbearable and illogical."

Answer: How the human being can act in the physical world depends entirely upon the physical instrumentality of his body. Higher ideas, for instance, can come to expression in this world only if there is a fully developed brain. Just as the pianist must wait until the piano builder has made a piano on which he can express his musical ideas, so does the soul have to wait with its faculties acquired in the previous life until the forces of the physical world have built up the bodily organs to the point where they can express these faculties. The nature forces have to go their way, the soul, also, has to go its way. To be sure, from the very beginning of human life a cooperation exists between soul and body forces. The soul works in the flexible and supple body of the child until it is made ready to become a bearer of the forces acquired in former life periods. For it is absolutely necessary that the reborn human being adjust himself to the new life conditions.

Were he simply to appear in a new life with all he has acquired previously, he would not fit into the surrounding world. For he has acquired his faculties and forces under quite different circumstances in completely different surroundings. Were he simply to enter the world in his former state he would be a stranger in it. The period of childhood is gone through in order to bring about harmony between the old and the new conditions. How would one of the cleverest ancient Romans appear in our present world, were he simply born into our world with his acquired powers? A power can only be employed when it is in harmony with the surrounding world. For instance, if a genius is born, the power of genius lies in the innermost being of this man which may be called the causal-body. The lower spirit-body and the body of feeling and sensation are adaptable, and in a certain sense not completely determined. These two parts of the human being are now elaborated. In this work the causal-body acts from within and the surroundings from without. With the completion of this work, these two parts may become the instruments of the acquired forces. — The thought that we have to be born as a child is, therefore, neither illogical nor unbearable. On the contrary, it would be unbearable were we born as a fully developed man into a world in which we are a stranger.

Question: "Are two successive incarnations of a human being similar to one another? Will an architect, for instance, become again an architect, a musician again a musician?"

Answer: This might be the case, but not necessarily so. Such similarities occur, but are by no means the rule. It is easy in this field to arrive at false conceptions because we form thoughts concerning the laws of reincarnation which cling too much to externalities. Someone loves the south, for instance, and therefore believes he must have been a southerner in a former incarnation. Such inclinations, however, do not reach up to the causal-body. They have a direct significance only for the one life. Whatever sends its effects over from one incarnation into another must be deeply seated in the central being of man. Let us assume, for instance, that someone is a musician in his present life. The spiritual harmonies and rhythms which express themselves in tones reach into the causal-body. The tones themselves belong to the outer physical life. They sit in the parts of the human being which come into existence and pass away. The lower ego or spirit-body, which is, at one time, the proper vehicle for tones may, in a subsequent life, be the vehicle for the perception of number and space relations. And the musician may now become a mathematician. Just through this fact the human being develops, in the course of his incarnations, into an all-comprehensive being by passing through the most manifold life activities. As has been stated, there are exceptions to this rule. And these are explicable by the great laws of the spiritual world.

Question: "What are the karmic facts in the case of a human being who is condemned to idiocy because of a defective brain?"

Answer: A case like this ought not to be dealt with by speculation and hypotheses, but only by means of spiritual-scientific experience. Therefore, the question here will be answered by quoting an example which has really occurred.

In a previous life a certain person had been doomed to an existence of mental torpor because of an undeveloped brain. During the time between his death and a new birth he was able to work over in himself all the depressing experiences of such a life, such as his having been pushed around, subjected to the unkindness of people, and he was reborn as a veritable genius of benevolence. Such a case shows clearly how wrong we can be if we refer everything in life karmicly back to the past. We cannot say in every instance: this destiny is the result of this or that guilt in the past. It is very well possible that an event has no relation whatsoever to the past but is only the cause for a karmic compensation in the future. An idiot need not have deserved his destiny through his deeds in the past. But the karmic consequence of his destiny for the future will not fail to appear. Just as a businessman's balance account is determined by the figures of his ledger, while he is free to have new receipts and expenses, so new deeds and blows of destiny may enter the life of a human being in spite of his book of life showing a definite balance at every given moment. Therefore, karma must not be conceived of as an immutable fate: it is absolutely compatible with the freedom, the will of man. Karma does not demand surrender to an unalterable fate; on the contrary, it affords us the certainty that no deed, no experience of the human being remains without effect or runs its course outside of the laws of the world. It affords us the certainty that every deed or experience is joined to just and compensating law. Moreover, if there were no karma, arbitrariness would rule in the world. As it is, I may know that every one of my actions, every one of my experiences is inserted in a lawful interrelationship. My deed is free; its effect follows definite laws. It is the free deed of a businessman when he makes a good deal; its result, however, shows up in the balance sheet of his ledger in accordance with definite laws.

How Karma Works

Sleep has often been called the younger brother of death. This simile illustrates the paths of the human spirit more exactly than a superficial observation might feel inclined to assume. For it gives us an idea of the way in which the most manifold incarnations passed through by this human spirit are interrelated. In the first chapter of this book, Reincarnation and Karma, Concepts Compelled by the Modern Scientific Point of View, it has been shown that the present natural-scientific mode of thought, if it but understands itself properly, leads to the ancient teaching of the evolution of the eternal human spirit through many lives. This knowledge is necessarily followed by the question: how are these manifold lives interrelated? In what sense is the life of a human being the effect of his former incarnations, and how does it become the cause of the later incarnations? The picture of sleep presents an image of the relation of cause and effect in this field. I arise in the morning. My continuous activity was interrupted during the night. I cannot resume this activity arbitrarily if order and connection are to govern my life. What I have done yesterday constitutes the conditions for my actions of today. I must make a connection with the result of my activities of yesterday. It is true in the fullest sense of the word that my deeds of yesterday are my destiny of today. I myself have shaped the causes to which I must add the effects. And I encounter these causes after having withdrawn from them for a short time. They belong to me, although I was separated from them for some time.

The effects of my experiences of yesterday belong to me in still another sense. I myself have been changed by them. Let us suppose that I have undertaken something in which I succeeded only partially. I have pondered on the reason for this partial failure. If I have again to carry out a similar task, I avoid the mistakes I have recognized. That is, I have acquired a new faculty. Thereby my experiences of yesterday have become the causes of my faculties of today. My past remains united with me; it lives on in my present; and it will follow me into my future. Through my past, I have created for myself the position in which I find myself at present. And the meaning of life demands that I remain united with this position. Would it not be senseless if, under normal conditions, I should not move into a house I had caused to be built for myself?

If the effects of my deeds of yesterday were not to be my destiny of today, I should not have to wake up today, but I should have to be created anew, out of the nothing. And the human spirit would have to be newly created, out of the nothing, if the results of its former lives were not to remain linked to its later lives. Indeed, the human being cannot live in any other position but the one which has been created through his previous life. He can do this no more than can certain animals, which have lost their power of sight as a result of their migration to the caves of Kentucky, live anywhere else but in these caves. They have, through their deed, through migration, created for themselves the conditions for their later existence. A being which has once been active is henceforth no longer isolated in the world; it has inserted itself into its deeds. And its future development is connected with what arises from the deeds. This connection of a being with the results of its deeds is the law of karma which rules the whole world. Activity that has become destiny is karma.

And sleep is a good picture of death for the reason that the human being, during sleep, is actually withdrawn from the field of action upon which destiny awaits him. While we sleep, the events on this field of action run their course. For a time, we have no influence upon this course. Nevertheless, we find again the effects of our actions, and we must link up with them. In reality, our personality every morning incarnates anew in our world of deeds. What was separated from us during the night, envelops us, as it were, during the day.

It is the same with the deeds of our former incarnations. Their results are embodied in the world in which we were incarnated. Yet they belong to us just as the life in the caves belongs to the animals which, through this life, have lost the power of sight. Just as these animals can only live if they find again the surroundings to which they have adapted themselves, so the human spirit is only able to live in those surroundings which, through his deeds, he has created for himself and are suited to him.

Every new morning the human body is ensouled anew, as it were. Natural science admits that this involves a process which it cannot grasp if it employs merely the laws it has gained in the physical world. Consider what the natural scientist Du Bois-Reymond says about this in his address, Die Grenze des Naturerkennens (The Limits of the Cognition of Nature): "If a brain, for some reason unconscious, as for instance in dreamless sleep, were to be viewed scientifically" — (Du Bois-Reymond says "astronomically") — "it would hold no longer any secrets, and if we were to add to this the natural-scientific knowledge of the rest of the body, there would be a complete deciphering of the entire human machine with its breathing, its heartbeat, its metabolism, its warmth, and so forth, right up to the nature of matter and force. The dreamless sleeper is comprehensible to the same degree that the world is comprehensible before consciousness appeared. But just as the world became doubly incomprehensible with the first stirring of consciousness, so the sleeper becomes incomprehensible with the first dream picture that arises in him." This cannot be otherwise. For, what the scientist describes here as the dreamless sleeper is that part of the human being which alone is subject to physical laws. The moment, however, it appears again permeated by the soul, it obeys the laws of the soul-life. During sleep, the human body obeys the physical laws: the moment the human being wakes up, the light of intelligent action flashes forth, like a spark, into purely physical existence. We speak entirely in the sense of the scientist Du Bois-Reymond when we state: the sleeping body may be investigated in all its aspects, yet we shall not be able to find the soul in it. But this soul continues the course of its rational deeds at the point where this was interrupted by sleep. — Thus the human being, also in this regard, belongs to two worlds. In one world he lives his bodily life which may be observed by means of physical laws; in the other he lives as a spiritual-rational being, and about this life we are able to learn nothing by means of physical laws. If we wish to study the bodily life, we have to hold to the physical laws of natural science; but if we wish to grasp the spiritual life, we have to acquaint ourselves with the laws of rational action, such, for instance, as logic, jurisprudence, economics, aesthetics, and so forth.

The sleeping human body, subject only to physical laws, can never accomplish anything in the realm of the laws of reason. But the human spirit carries these laws of reason into the physical world. And just as much as he has carried into it will he find again when, after an interruption, he resumes the thread of his activity.

Let us hold on to the picture of sleep. If life is not to be meaningless, the personality has to link up today with its deeds of yesterday. It could not do so did it not feel itself joined to these deeds. I should be unable to pick up today the result of my activity of yesterday, had there not remained within myself something of this activity. If I had today forgotten everything that I have experienced yesterday, I should be a new human being, unable to link up with anything. It is my memory which enables me to link up with my deeds of yesterday. — This memory binds me to the effects of my action. That which, in the real sense, belongs to my life of reason, — logic, for instance, — is today the same it was yesterday. This is applicable also to that which did not enter my field of vision yesterday, indeed, which never entered it. My memory connects my logical action of today with my logical action of yesterday. If matters depended merely upon logic, we certainly might start a new life every morning. But memory retains what binds us to our destiny.

Thus I really find myself in the morning as a threefold being. I find my body again which during my sleep has obeyed its merely physical laws. I find again my own self, my human spirit, which is today the same it was yesterday, and which is today endowed with the gift of rational action with which it was endowed yesterday. And I find — preserved by memory — everything that my yesterday, that my entire past has made of me. —

And this affords us at the same time a picture of the threefold being of man. In every new incarnation the human being finds himself in a physical organism which is subject to the laws of external nature. And in every incarnation he is the same human spirit. As such he is the Eternal within the manifold incarnations. Body and Spirit confront one another. Between these two there must lie something just as memory lies between my deeds of yesterday and those of today. And this something is the soul. It preserves the effects of my deeds from former lives and brings it about that the spirit, in a new incarnation, appears in the form which previous earth lives have given it. In this way, body, soul, and spirit are interrelated. The spirit is eternal; birth and death rule in the body according to the laws of the physical world; both are brought together again and again by the soul as it fashions our destiny out of our deeds. (Each of the above-mentioned principles: body, soul, and spirit, in turn consists of three members. Thus the human being appears to be formed of nine members. The body consists of: (1) the actual body, (2) the life-body, (3) the sentient-body. The soul consists of: (4) the sentient-soul, (5) the intellectual-soul, (6) the consciousness-soul. The spirit consists of: (7) spirit-self, (8) life-spirit, (9) spirit-man. In the incarnated human being, 3 and 4, and 6 and 7 unite, flowing into one another. Through this fact the nine members appear to have contracted into seven members.)

In regard to the comparison of the soul with memory we are also in a position to refer to modern natural science. The scientist Ewald Hering published a treatise in 1870 which bears the title: Ueber das Gedaechtnis als eine allgemeine Funktion der organisierten Materie (Memory as a General Function of Organized Matter). Ernst Haeckel agrees with Hering's point of view. He states the following in his treatise: Ueber die Wellenzeugung der Lebensteilchen (The Wave Generation of Living Particles): "Profound reflection must bring the conviction that without the assumption of an unconscious memory of living matter the most important life functions are utterly inexplicable. The faculty of forming ideas and concepts, of thinking and consciousness, of practice and habit, of nutrition and reproduction rests upon the function of the unconscious memory, the activity of which is much more significant than that of conscious memory. Hering is right in stating that it is memory to which we owe nearly everything that we are and have." And now Haeckel tries to trace back the processes of heredity within living creatures to this unconscious memory. The fact that the daughter-being resembles the mother-being, that the former inherits the qualities of the latter, is thus supposed to be due to the unconscious memory of the living, which in the course of reproduction retains the memory of the preceding forms. — It is not a question here of investigating how much of the presentations of Hering and Haeckel are scientifically tenable; for our purposes it suffices to draw attention to the fact that the natural scientist is compelled to assume an entity which he considers similar to memory; he is compelled to do so if he goes beyond birth and death, and presumes something that endures beyond death. He quite naturally seizes upon a supersensible force in the realm where the laws of physical nature do not suffice.

We must, however, realize that we are dealing here merely with a comparison, with a picture, when we speak of memory. We must not believe that by soul we understand something that is equivalent to conscious memory. Even in ordinary life it is not always conscious memory that is active when we make use of the experiences of the past. We bear within us the fruits of these experiences even if we do not always consciously remember what we have experienced. Who can remember all the details of his learning to read and write? Moreover, who was ever conscious of all those details? Habit, for instance, is a kind of unconscious memory. — By means of this comparison with memory we merely wish to point to the soul which inserts itself between body and spirit and constitutes the mediator between the Eternal and that which, as the Physical, is inwoven into the course of birth and death.

The spirit that reincarnates thus finds within the physical world the results of its deeds as its destiny; and the soul that is bound to it, mediates the spirit's linking up with this destiny. Now we may ask: how can the spirit find the results of its deeds, since, on reincarnating, it is certainly placed in a world completely different from the one in which it existed previously? This question is based upon a very externalized conception of the web of destiny. If I transfer my residence from Europe to America, I, too, find myself in completely new surroundings. Yet my life in America is completely dependent upon my previous life in Europe. If I have been a mechanic in Europe, my life in America will take on a form quite different from the one it would take on had I been a bank clerk. In the one case I shall probably be surrounded in America by machines, in the other by banking papers. In every case my previous life determines my surroundings, it attracts, as it were, out of the whole environment those things which are related to it. This is also the case with my spirit-soul. It surrounds itself quite necessarily with what it is related to out of its previous life. This cannot constitute a contradiction of the simile of sleep and death if we realize that we are dealing only with a simile, although a most striking one. That I find in the morning the situation which I myself have created on the previous day is brought about by the direct course of events. That I find on reincarnating an environment that corresponds to the result of my deeds of the previous life is brought about through the affinity of my reborn spirit-soul with the things of this environment.

What leads me into this environment? Directly the qualities of my spirit-soul on reincarnating. But I possess these qualities merely through the fact that the deeds of my previous lives have implanted them into the spirit-soul. These deeds, therefore, are the real cause of my being born into certain circumstances. And what I do today will be one of the causes of my finding myself in a later life within certain definite circumstances. — Thus man indeed creates his destiny for himself. This remains incomprehensible only as long as one considers the separate life as such and dos not regard it as a link in the chain of successive lives.

Thus we may say that nothing can happen to the human being in life for which he has not himself created the conditions. Only through insight into the law of destiny — karma — does it become comprehensible why "the good man has often to suffer, while the evil one may experience happiness." This seeming disharmony of the one life disappears when the view is extended upon many lives. — To be sure, the law of karma must not be conceived of as being so simple that we might compare it to an ordinary judge or to civil justice. This would be the same as if we were to imagine God as an old man with a white beard. Many people fall into this error. Especially the opponents of the idea of karma proceed from such erroneous premises. They fight against the conception which they impute to the believers in karma and not against the conception held by the true knowers.

What is the relation of the human being to his physical surroundings when he enters a new incarnation? This relation is composed of two factors: first, in the time between two consecutive incarnations he has had no part in the physical world; second, he passed through a certain development during that period. It is self-evident that no influence from the physical world can affect this development, for the spirit-soul then exists outside this physical world. Everything that takes place in the spirit-soul, it can, therefore, only draw out of itself, that is to say, out of the super-physical world. During its incarnation it was interwoven with the physical world of facts; after its discarnation through death, it is deprived of the direct influence of this factual world. It has merely retained from the latter that which we have compared to memory. — This "memory remnant" consists of two parts. These parts become evident if we consider what has contributed to its formation. — The spirit has lived in the body and through the body, therefore, it entered into relation with the bodily surroundings. This relation has found its expression through the fact that, by means of the body, impulses, desires, and passions have developed and that, through them, outer actions have been performed. Because he has a corporeal existence, the human being acts under the influence of impulses, desires, and passions. And these have a significance in two directions. On the one hand, they impress themselves upon the outer actions which the human being performs. And on the other, they form his personal character. The action I perform is the result of my desire; and I myself, as a personality, am what is expressed by this desire. The action passes over into the outer world;the desire remains within my soul just as the thought remains within my memory. And just as the thought image in my memory is strengthened through every new impression of like nature, so is the desire strengthened through every new action which I perform under its influence. Thus within my soul, because of corporeal existence, there lives a certain sum of impulses, desires, and passions. The sum total of these is designated by the expression "body of desire." — This body of desire is intimately connected with physical existence, for it comes into being under the influence of the physical corporeality. The moment the spirit is no longer incarnated it cannot continue the formation of this body of desire. The spirit must free itself from this desire-body in so far as it was connected, through it, with the single physical life. The physical life is followed by another in which this liberation occurs. We may ask: Does not death signify the destruction also of this body of desire? The answer is: No; for to the degree in which, at every moment of physical life, desire surpasses satisfaction, desire persists even when the possibility of satisfaction has ceased. Only a human being who does not desire anything of the physical world has no surplus of desire over satisfaction. Only a man of no desires dies without retaining in his spirit a certain amount of desire. And this amount must gradually diminish and fade away after death. The state of this fading away is called "the sojourn in the region of desire." It can easily be seen that the more the human being has felt bound to the sense life, the longer must this state persist.

The second part of the "memory remnant" is formed in a different way. Just as desire draws the spirit toward the past life, so this second part directs it toward the future. The spirit, through its activity in the body, has become acquainted with the world to which this body belongs. Each new exertion, each new experience enhances this acquaintance. As a rule the human being does a thing better the second time than he does it the first. Experience impresses itself upon the spirit, enhancing its capacities. Thus our experience acts upon our future, and if we have no longer the opportunity to have experiences, then the result of these experiences remains as memory remnant. — But no experience could affect us if we did not have the capacity to make use of it. The way in which we are able to absorb the experience, the use we are able to make of it, determines its significance for our future. For Goethe, an experience had a significance quite different from the significance it had for his valet; and it produced results for Goethe quite different from those it produced for his valet. What faculties we acquire through an experience depends, therefore, upon the spiritual work we perform in connection with the experience. — I always have within me, at any given moment of my life, a sum total of the results of my experience. And this sum total forms the potential of capacities which may appear in due course. — Such a sum total of experiences the human spirit possesses when it discarnates. This the human spirit takes with it into supersensible life. Now, when it is no longer bound to physical existence by bodily ties and when it has divested itself also of the desires which chain it to this physical existence, then the fruit of its experience has remained with the spirit. And this fruit is completely freed from the direct influence of the past life. The spirit can now devote itself entirely to what it is capable of fashioning out of this fruit for the future. Thus the spirit, after having left the region of desire, is in a state in which its experiences of former lives transform themselves into potentials — that is to say, talents, capacities — for the future. The life of the spirit in this state is designated as the sojourn in the "region of bliss." ("Bliss" may, indeed, designate a state in which all worry about the past is relegated to oblivion and which permits the heart to beat solely for the concerns of the future.) It is self-evident that the greater the potentiality exists at death for the acquirement of new capacities, the longer will this state in general last.

Naturally, it cannot be a question here of developing the complete scope of knowledge relating to the human spirit. We merely intend to show how the law of karma operates in physical life. For this purpose it is sufficient to know what the spirit takes out of this physical life into supersensible states and what it brings back again for a new incarnation. It brings with it the results of the experiences undergone in previous lives, transformed into the capacities of its being. — In order to realize the far-reaching character of this fact we need only elucidate the process by a single example. The philosopher, Kant, says: "Two things fill the soul with ever increasing wonder: the starry heavens above me and the moral law within me." Every thinking human being must admit that the starry heavens have not sprung out of nothingness but have come gradually into existence. And it is Kant himself who in 1755, in a basic treatise, tried to explain the gradual formation of a cosmos. Likewise, however, we must not accept the fact of moral law without an explanation. This moral law, too, has not sprung from nothingness. In the first incarnations through which man passed the moral law did not speak in him in the way it spoke in Kant. Primitive man acts in accordance with his desires. And he carries the experiences which he has undergone through such action into the supersensible states. Here they become higher faculties. And in a subsequent incarnation, mere desire no longer acts in him, but it is now guided by the effect of the previous experiences. And many incarnations are needed before the human being, originally completely given over to desires, confronts the surrounding world with the purified moral law which Kant designates as something demanding the same admiration as is demanded by the starry heavens.

The surrounding world into which the human being is born through a new incarnation confronts him with the results of his deeds, as his destiny. He himself enters this surrounding world with the capacities which he has fashioned for himself in the supersensible state out of his former experiences. Therefore his experiences in the physical world will, in general, be at a higher level the more often he has incarnated, or the greater his efforts were during his previous incarnations. Thus his pilgrimage through the incarnations will be an upward development. The treasure which his experiences accumulate in his spirit will become richer and richer. And he thereby confronts his surrounding world, his destiny, with greater and greater maturity. This makes him increasingly the master of his destiny. For what he gains through his experiences is the fact that he learns to grasp the laws of the world in which these experiences occur. At first the spirit does not find its way about in the surrounding world. It gropes in the dark. But with every new incarnation the world grows brighter. The spirit acquires a knowledge of the laws of its surrounding world; in other words, it accomplishes ever more consciously what it previously did in dullness of mind. The compulsion of the surrounding world decreases; the spirit becomes increasingly self-determinative. The spirit, however, which is self-determinative, is the free spirit. Action in the full clear light of consciousness is free action. (I have tried to present the nature of the free human spirit in my book, Philosophie der Freiheit, (Philosophy of Freedom — Spiritual Activity.) The full freedom of the human spirit is the ideal of its development. We cannot ask the question: is man free or unfree? The philosophers who put the question of freedom in this fashion can never acquire a clear thought about it. For the human being in his present state is neither free nor unfree; but he is on the way to freedom. He is partially free, partially unfree. He is free to the degree he has acquired knowledge and consciousness of world relations. — The fact that our destiny, our karma, meets us in the form of absolute necessity is no obstacle to our freedom. For when we act we approach this destiny with the measure of independence we have achieved. It is not destiny that acts, but it is we who act in accordance with the laws of this destiny.

If I light a match, fire arises according to necessary laws;but it was I who put these necessary laws into effect. Likewise, I can perform an action only in the sense of the necessary laws of my karma, but it is I who puts these necessary laws into effect. And new karma is created through the deed proceeding from me, just as the fire, according to necessary laws of nature, continues to be effective after I have kindled it.

This also throws light upon another doubt which may assail a person in regard to the effectiveness of the law of karma. Somebody might say: "If karma is an unalterable law, then it is wrong to help a person. For what befalls him is the consequence of his karma, and it is absolutely necessary that it should befall him." Certainly, I cannot eliminate the effects of the destiny which a human spirit has created for himself in former incarnations. But the matter of importance here is how he finds his way into this destiny, and what new destiny he may create for himself under the influence of the old one. If I help him, I may bring about the possibility of his giving his destiny a favorable turn through his deeds; if I refrain from helping him, the opposite may perhaps occur. Naturally, everything will depend upon whether my help is a wise or unwise one. [The fact that I am present to help may be a part of both his Karma and mine, or my presence and deed may be a free act.]

His advance through ever new incarnations signifies a higher development of the human spirit. This higher development comes to expression in the fact that the world in which the incarnations of the spirit take place is comprehended in increasing measure by this spirit. This world, however, comprises the incarnations themselves. In regard to the latter, too, the spirit gradually passes from a state of unconsciousness to one of consciousness. On the path of evolution there lies the point from which the human being is able to look back upon his successive incarnations with full consciousness. — This is a thought at which it is easy to mock; and it is easy to criticise it negatively. But whoever does this has no idea of the nature of such truths. And derision as well as criticism place themselves like a dragon in front of the portal of the sanctuary within which we may attain knowledge of these truths. For it is self-evident that truths, the realization of which lies for the human being in the future, cannot be found as facts in the present. There is only one way of convincing oneself of their reality: namely, to make every effort possible to attain this reality.

Reincarnation and Karma

Francesco Redi, the Italian natural scientist, was considered a dangerous heretic by the leading scholars of the seventeenth century because he maintained that even the lowest animals originate through reproduction. He narrowly escaped the martyr-destiny of Giordano Bruno or Galileo. For the orthodox scientist of that time believed that worms, insects, and even fish could originate out of lifeless mud. Redi maintained that which today is generally acknowledged: that all living creatures have descended from living creatures. He committed the sin of recognizing a truth two centuries before science found its "irrefutable" proof. Since Pasteur has carried out his investigations, there can be no longer any doubt about the fact that those cases were merely illusion in which people believed that living creatures could come into existence out of lifeless substances through "spontaneous generation". The life germs entering such lifeless substances escaped observation. With proper means, Pasteur prevented the entrance of such germs into substances in which, ordinarily, small living creatures come into existence, and not even a trace of the living was formed. Thus it was demonstrated that the living springs only from the life germ. Redi had been completely correct.

Today, the spiritual scientist, the anthroposophist, finds himself in a situation similar to that of the Italian scientist.

On the basis of his knowledge, he must maintain in regard to the soul what Redi maintained in regard to life. He must maintain that the soul nature can spring only from the soul. And if science advances in the direction it has taken since the seventeenth century, then the time will come when, out of its own nature, science will uphold this view. For — and this must be emphasized again and again — the attitude of thought which underlies the anthroposophical conception of today is no other than the one underlying the scientific dictum that insects, worms and fish originate from life germs and not from mud. The anthroposophical conception maintains the postulate: "Every soul originates out of the soul nature," in the same sense and with the same significance in which the scientist maintains: "Everything living originates out of the living."

Today's customs differ from those of the seventeenth century. The attitudes of mind underlying the customs have not changed particularly. To be sure, in the seventeenth century, heretical views were persecuted by means no longer considered human today. Today, spiritual scientists, anthroposophists, will not be threatened with burning at the stake: one is satisfied in rendering them harmless by branding them as visionaries and unclear thinkers. Current science designates them fools. The former execution through the inquisition has been replaced by modern, journalistic execution. The anthroposophists, however, remain steadfast; they console themselves in the consciousness that the time will come when some Virchow will say: "There was a time — fortunately it is now superceded — when people believed that the soul comes into existence by itself if certain complicated chemical and physical processes take place within the skull. Today, for every serious researcher this infantile conception must give way to the statement that everything pertaining to the soul springs from the soul."

One must by no means believe that spiritual science intends to prove its truths through natural science. It must be emphasized, however, that spiritual science has an attitude of mind similar to that of true natural science. The anthroposophist accomplishes in the sphere of the soul life what the nature researcher strives to attain in the domains perceptible to the eyes and audible to the ears. There can be no contradiction between genuine natural science and spiritual science. The anthroposophist demonstrates that the laws which he postulates for the soul life are correspondingly valid also for the external phenomena of nature. He does so because he knows that the human sense of knowledge can only feel satisfied if it perceives that harmony, and not discord, rules among the various phenomenal realms of existence. Today most human beings who strive at all for knowledge and truth are acquainted with certain natural-scientific conceptions. Such truths can be acquired, so to speak, with the greatest ease. The science sections of newspapers disclose to the educated and uneducated alike the laws according to which the perfect animals develop out of the imperfect, they disclose the profound relationship between man and the anthropoid ape, and smart magazine writers never tire of inculcating their readers with their conception of "spirit" in the age of the "great Darwin." They very seldom add that in Darwin's main treatise there is to be found the statement: "I hold that all organic beings that have ever lived on this earth have descended from one primordial form into which the creator breathed the breath of life." (Origin of Species, Vol. II, chapter XV.) — In our age it is most important to show again and again that Anthroposophy does not treat the conceptions of "the breathing in of life" and the soul as lightly as Darwin and many a Darwinian, but that its truths do not contradict the findings of true nature research. Anthroposophy does not wish to penetrate into the mysteries of spirit-life upon the crutches of natural science of the present age, but it merely wishes to say: "Recognize the laws of the spiritual life and you will find these sublime laws verified in corresponding form if you descend to the domain in which you can see with eyes and hear with ears." Natural science of the present age does not contradict spiritual science; on the contrary, it is itself elemental spiritual science. Only because Haeckel applied to the evolution of animal life the laws which the psychologists since ancient days have applied to the soul, did he achieve such beautiful results in the field of animal life. If he himself is not of this conviction, it does not matter; he simply does not know the laws of the soul, nor is he acquainted with the research which can be carried on in the field of the soul. The significance of his findings in his field is thereby not diminished. Great men have the faults of their virtues. Our task is to show that Haeckel in the field where he is competent is nothing but an anthroposophist. — By linking up with the natural-scientific knowledge of the present age, still another aid offers itself to the spiritual scientist. The objects of outer nature are, so to speak, to be grasped by our hands. It is, therefore, easy to expound their laws. It is not difficult to realize that plants change when they are transplanted from one region into another. Nor is it hard to visualize that a certain animal species loses its power of eyesight when it lives for a certain length of time in dark caves. By demonstrating the laws which are active in such processes, it is easy to lead over to the less manifest, less comprehensible laws which we encounter in the field of the soul life. — if the anthroposophist employs natural science as an aid, he merely does so in order to illustrate what he is saying. He has to show that anthroposophic truths, with respective modifications, are to be found in the domain of natural science, and that natural science cannot be anything but elemental spiritual science; and he has to employ natural-scientific concepts in order to lead over to his concepts of a higher nature.

The objection might be raised here that any inclination toward present-day natural-scientific conceptions might put spiritual science into an awkward position for the simple reason that these conceptions themselves rest upon a completely uncertain foundation. It is true: There are scientists who consider certain fundamental principles of Darwinism as irrefutable, and there are others who even today speak of a "crisis in Darwinism." The former consider the concepts of "the omnipotence of natural selection" and "the struggle for survival" to be a comprehensive explanation of the evolution of living creatures; the latter consider this "struggle for survival" to be one of the infantile complaints of modern science and speak of the "impotence of natural selection." — If matters depended upon these specific, problematic questions, it were certainly better for the anthroposophist to pay no attention to them and to wait for a more propitious moment when an agreement with natural science might be achieved. But matters do not depend upon these problems. What is important, however, is a certain attitude, a mode of thought within natural-scientific research in our age, certain definite great guiding lines, which are adhered to everywhere, even though the thoughts of various researchers and thinkers concerning specific questions diverge widely. It is true: Ernst Haeckel's and Virchow's conceptions of the "genesis of man" diverge greatly. But the anthroposophical thinker might consider himself fortunate if leading personalities were to think as clearly about certain comprehensive viewpoints concerning the soul life as these opponents think about that which they consider absolutely certain in spite of their disagreement. Neither the adherents of Haeckel nor those of Virchow search today for the origin of worms in lifeless mud; neither the former nor the latter doubt that "all living creatures originate from the living," in the sense designated above. — In psychology we have not yet advanced so far. Clarity is completely lacking concerning a view point which might be compared with such scientific fundamental convictions. Whoever wishes to explain the shape and mode of life of a worm knows that he has to consider its ovum and ancestors; he knows the direction in which his research must proceed, although the viewpoints may differ concerning other aspects of the question, or even the statement may be made that the time is not yet ripe when definite thoughts may be formed concerning this or that point. — Where, in psychology, is there to be found a similar clarity? The fact that the soul has spiritual qualities, just as the worm has physical ones, does not cause the researcher to approach — as he should — the one fact with the same attitude of mind as he approaches the other. To be sure, our age is under the influence of thought habits which prevent innumerable people, occupied with these problems, from entering at all properly upon such demands. — True, it will be admitted that the soul qualities of a human being must originate somewhere just as do the physical ones. The reasons are being sought for the fact that the souls of a group of children are so different from one another, although the children all grew up and were educated under identical circumstances; that even twins differ from one another in essential characteristics, although they always lived at the same place and under the care of the same nurse. The case of the Siamese Twins is quoted, whose final years of life were, allegedly, spent in great discomfort in consequence of their opposite sympathies concerning the North-American Civil War. We do not deny that careful thought and observation have been directed upon such phenomena and that remarkable studies have been made and results achieved. But the fact remains that these efforts concerning the soul life are on a par with the efforts of a scientist who maintains that living creatures originate from lifeless mud. In order to explain the lower psychic qualities, we are undoubtedly justified in pointing to the physical forebears and in speaking of heredity, just as we do in the case of bodily traits. But we deliberately close our eyes to the most important aspect of the matter if we proceed in the same direction with respect to the higher soul qualities, the actually spiritual in man. We have become accustomed to regard these higher soul qualities as a mere enhancement, as a higher degree of the lower ones. And we therefore believe that an explanation might satisfy us which follows the same lines as the explanation offered for the soul qualities of the animal.

It is not to be denied that the observation of certain soul functions of higher animals may easily lead to this mistaken conception. We only need draw attention to the fact that dogs show remarkable proof of a faithful memory; that horses, noticing the loss of a horse shoe, walk of their own accord to the blacksmith who has shod them before; that animals which are shut up in a room, can by themselves open the door; we might quote many more of these astonishing facts. Certainly, the anthroposophist, too, will not refrain from admitting the possibility of continued enhancement of animal faculties. But must we, for that reason, obliterate the difference between the lower soul traits which man shares with the animal, and the higher spiritual qualities which man alone possesses? This can only be done by someone who is completely blinded by the dogmatic prejudice of a "science" which wishes to stick fast to the facts of the coarse, physical senses. Simply consider what is established by indisputable observation, namely, that animals, even the highest-developed ones, cannot count and therefore are unable to learn arithmetic. The fact that the human being is distinguished from the animal by his ability to count was considered a significant insight even in ancient schools of wisdom. — Counting is the simplest, the most insignificant of the higher soul faculties. For that very reason we cite it here, because it indicates the point where the animal-soul element passes over into the spirit-soul element, into the higher human element. Of course, it is very easy to raise objections here also. First, one might say that we have not yet reached the end of the world and that we might one day succeed in what we have not yet been able to do, namely, to teach counting to intelligent animals. And secondly, one might point to the fact that the brain has reached a higher stage of perfection in man than in the animal, and that herein lies the reason for the human brain's higher degrees of soul activity. We may fully concur with the persons who raise these objections. Yet we are in the same position concerning those people who, in regard to the fact that all living creatures spring from the living, maintain over and over again that the worm is governed by the same chemical and physical laws that govern the mud, only in a more complicated manner. Nothing can be done for a person who wishes to disclose the secrets of nature by means of trivialities and what is self-evident. There are people who consider the degree of insight they have attained to be the most penetrating imaginable and to whom, therefore, it never occurs that there might be someone else able to raise the same trivial objections, did he not see their worthlessness. — No objection can be raised against the conception that all higher processes in the world are merely higher degrees of the lower processes to be found in the mud. But just as it is impossible for a person of insight today to maintain that the worm originates from the mud, so is it impossible for a clear thinker to force the spirit-soul nature into the same concept-pattern as that of the animal-soul nature. Just as we remain within the sphere of the living in order to explain the descent of the living, so must we remain in the sphere of the soul-spirit nature in order to understand the soul-spirit nature's origin.

There are facts which may be observed everywhere and which are bypassed by countless people without their paying any attention to them. Then someone appears who, by becoming aware of one of these facts, discovers a fundamental and far-reaching truth. It is reported that Galileo discovered the important law of the pendulum by observing a swinging chandelier in the cathedral of Pisa. Up to that time, innumerable people had seen swinging church lamps without making this decisive observation. What matters in such cases is that we connect the right thoughts with the things we see. Now, there exists a fact which is quite generally accessible and which, when viewed in an appropriate manner, throws a clear light upon the character of the soul-spirit nature. This is the simple truth that every human being has a biography, but not the animal. To be sure, certain people will say: Is it not possible to write the life story of a cat or a dog? The answer must be: Undoubtedly it is; but there is also a kind of school exercise which requires the children to describe the fate of a pen. The important point here is that the biography has the same fundamental significance in regard to the individual human being as the description of the species has in regard to the animal. Just as I am interested in the description of the lion-species in regard to the lion, so am I interested in the biography in regard to the individual human being. By describing their human species, I have not exhaustively described Schiller, Goethe, and Heine, as would be the case regarding the single lion once I have recognized it as a member of its species. The individual human being is more than a member of his species. Like the animal, he shares the characteristics of his species with his physical forebears. But where these characteristics terminate, there begins for the human being his unique position, his task in the world. And where this begins, all possibility of an explanation according to the pattern of animal-physical heredity ceases. I may trace back Schiller's nose and hair, perhaps even certain characteristics of his temperament, to corresponding traits in his ancestors, but never his genius. And naturally, this does not only hold good for Schiller. This also holds good for Mrs. Miller of Gotham. In her case also, if we are but willing, we shall find soul-spiritual characteristics which cannot be traced back to her parents and grand-parents in the same way we can trace the shape of her nose or the blue color of her eyes. It is true, Goethe has said that he had received from his father his figure and his serious conduct of life, and from his little mother his joyous nature and power of fantasy, and that, as a consequence, nothing original was to be found in the whole man. But in spite of this, nobody will try to trace back Goethe's gifts to father and mother — and be satisfied with it — in the same sense in which we trace back the form and manner of life of the lion to his forebears. — This is the direction in which psychology must proceed if it wishes to parallel the natural-scientific postulate that "all living creatures originate from the living" with the corresponding postulate that "everything of the nature of the soul is to be explained by the soul-nature." We intend to follow up this direction and show how the laws of reincarnation and karma, seen from this point of view, are a natural-scientific necessity. It seems most peculiar that so many people pass by the question of the origin of the soul-nature simply because they fear that they might find themselves caught in an uncertain field of knowledge. They will be shown what the great scientist Carl Gegenbaur has said about Darwinism. Even if the direct assertions of Darwin may not be entirely correct, yet they have led to discoveries which without them would not have been made. In a convincing manner Darwin has pointed to the evolution of one form of life out of another one, and this has stimulated the research into the relationships of such forms. Even those who contest the errors of Darwinism ought to realize that this same Darwinism has brought clarity and certainty to the research into animal and plant evolution, thus throwing light into dark reaches of the working of nature. Its errors will be overcome by itself. If it did not exist, we should not have its beneficial consequences. In regard to the spiritual life, the person who fears uncertainty concerning the anthroposophical conception ought to concede to it the same possibility; even though anthroposophical teachings were not completely correct, yet they would, out of their very nature, lead to the light concerning the riddles of the soul. To them, too, we shall owe clarity and certainty. And since they are concerned with our spiritual destiny, our human destination, our highest tasks, the bringing about of this clarity and certainty ought to be the most significant concern of our life. In this sphere, striving for knowledge is at the same time a moral necessity, an absolute moral duty.

David Friedrich Strauss endeavored to furnish a kind of Bible for the "enlightened" human being in his book, Der alte und neue Glaube (Faith — Ancient and Modern). "Modern faith" is to be based on the revelations of natural science, and not on the revelations of "ancient faith" which, in the opinion of this apostle of enlightenment, have been superceded. This new Bible has been written under the impression of Darwinism by a personality who says to himself: Whoever, like myself, counts himself among the enlightened, has ceased, long before Darwin, to believe in "supernatural revelation" and its miracles. He has made it clear to himself that in nature there hold sway necessary, immutable laws, and whatever miracles are reported in the Bible would be disturbances, interruptions of these laws; and there cannot be such disturbances and interruptions. We know from the laws of nature that the dead cannot be reawakened to life: therefore, Jesus cannot have reawakened Lazarus. — However, — so this enlightened person continues — there was a gap in our explanation of nature. We were able to understand how the phenomena of the lifeless may be explained through immutable laws of nature; but we were unable to form a natural conception about the origin of the manifold species of plants and animals and of the human being himself. To be sure, we believed that in their case also we are concerned merely with necessary laws of nature; but we did not know their nature nor their mode of action. Try as we might, we were unable to raise reasonable objection to the statement of Carl von Linné, the great nature-researcher of the eighteenth century, that there exist as many "species in the animal and plant kingdom as were originally created in principle." Were we not confronted here with as many miracles of creation as with species of plants and animals? Of what use was our conviction that God was unable to raise Lazarus through a supernatural interference with the natural order, through a miracle, when we had to assume the existence of such supernatural deeds in countless numbers. Then Darwin appeared and showed us that, through immutable laws of nature (natural selection and struggle for life), the plant and animal species come into existence just as do the lifeless phenomena. Our gap in the explanation of nature was filled.

Out of the mood which this conviction engendered in him, David Friedrich Strauss wrote down the following statement of his "ancient and modern belief": "We philosophers and critical theologians spoke to no purpose in denying the existence of miracles; our authoritative decree faded away without effect because we were unable to prove their dispensability and give evidence of a nature force which could replace them in the fields where up to now they were deemed most indispensable. Darwin has given proof of this nature force, this nature process, he has opened the door through which a fortunate posterity will cast the miracle into oblivion. Everybody who knows what is connected with the concept 'miracle' will praise him as one of the greatest benefactors of the human race."

These words express the mood of the victor. And all those who feel like Strauss may disclose the following view of the "modern faith": Once upon a time, lifeless particles of matter have conglomerated through their inherent forces in such a way as to produce living matter. This living matter developed, according to necessary laws, into the simplest, most imperfect living creatures. These, according to similarly necessary laws, transformed themselves further into the worm, the fish, the snake, the marsupial, and finally into the ape. And since Huxley, the great English nature researcher, has demonstrated that human beings are more similar in their structure to the most highly developed apes than the latter are to the lower apes, what then stands in the way of the assumption that the human being himself has, according to the same natural laws, developed from the higher apes? And further, do we not find what we call higher human spiritual activity, what we call morals, in an imperfect condition already with the animal. May we doubt the fact that the animals — as their structure became more perfect, as it developed into the human form, merely on the basis of physical laws — likewise developed the indications of intellect and morals to be found in them to the human stage?

All this seems to be perfectly correct. Although everybody must admit that our knowledge of nature will not for a long time to come be in the position to conceive of how what has been described above takes place in detail, yet we shall discover more and more facts and laws; and thus the "modern faith" will gain more and firmer supports.

Now it is a fact that the research and study of recent years have not furnished such solid supports for this belief; on the contrary, they have contributed greatly to discredit it. Yet it holds sway in ever extending circles and is a great obstacle to every other conviction.

There is no doubt that if David Friedrich Strauss and those of like mind are right, then all talk of higher spiritual laws of existence is an absurdity; the "modern faith" would have to be based solely on the foundations which these personalities assert are the result of the knowledge of nature.

Yet, whoever with unprejudiced mind follows up the statements of these adherents of the "modern faith" is confronted by a peculiar fact. And this fact presses upon us most irresistibly if we look at the thoughts of those people who have preserved some degree of impartiality in the face of the self-assured assertions of these orthodox pioneers of progress.

For there are hidden corners in the creed of these modern believers. And if we uncover what exists in these corners, then the true findings of modern natural science shine forth in full brilliance, but the opinions of the modern believers concerning the human being begin to fade away.

Let us throw light into a few of these corners. At the outset, let us keep to that personality who is the most significant and the most venerable of these modern believers. On page 804 of the ninth edition of Haeckel's Natuerliche Schoepfungsgeschichte (Natural Genesis) we read: "The final result of a comparison of animals and man shows that between the most highly developed animal souls and the lowest human souls there exists only a small quantitative, but no qualitative difference; this difference is much smaller than the difference between the lowest and the highest human souls, or the difference between the highest and the lowest animal souls." Now, what is the modern believer's attitude toward such a fact? He announces: we must explain the difference between the lower and the higher animal souls as a consequence of necessary and immutable laws. And we study these laws. We ask ourselves: how did it come about that out of animals with a lower soul have developed those with a higher soul? We look in nature for conditions through which the lower may develop into the higher. We then find, for example, that animals which have migrated to the caves of Kentucky become blind there. It becomes clear to us that through the sojourn in the darkness the eyes have lost their function. In these eyes the physical and chemical processes no longer take place which were carried out during the act of seeing. The stream of nourishment which has formerly been used for this activity is now diverted to other organs. The animals change their shape. In this way, new animal species can arise out of existing ones if only the transformation which nature causes in these species is sufficiently great and manifold. — What actually takes place here? Nature brings about changes in certain beings; and these changes later also appear in their descendants. We say: they are transmitted by heredity. Thus the coming into existence of new animal and plant species is explained.

The modern believers now continue happily in the direction of their explanation. The difference between the lowest human souls and the highest animal souls is not particularly great. Therefore, certain life conditions in which the higher animal souls have been placed have brought about changes by means of which they became lower human souls. The miracle of the evolution of the human soul has been cast out of the temple of the "modern faith" into oblivion, to use an expression of Strauss', and man has been classified among the animals according to "eternal, necessary" laws. Satisfied, the modern believer retires into peaceful slumber; he does not wish to go further.

Honest thinking must disturb his slumber. For this honest thinking must keep alive around his couch the spirits which he himself has evoked. Let us consider more closely the above statement of Haeckel: "the difference (between higher animals and men) is much smaller than the difference between the lowest and the highest human souls." If the modern believer admits this, may he then indulge in peaceful slumber as soon as he — according to his opinion — has explained the evolution of the lower men out of the highest animals?

No, he must not do this, and if he does so nevertheless, then he denies the whole basis upon which he has founded his conviction. What would a modern believer reply to another who were to say: I have demonstrated how fish have originated from lower living creatures. This suffices. I have shown that everything evolves — therefore the species higher than the fish will doubtless have developed like the fish. There is no doubt that the modern believer would reply: Your general thought of evolution is useless; you must be able to show how the mammals originate; for there is a greater difference between mammals and fish than between fish and those animals on a stage directly below them. — And what would have to be the consequence of the modern believer's real faithfulness to his creed? He would have to say: the difference between the higher and lower human souls is greater than the difference between these lower souls and the animal souls on the stage directly below them; therefore I must admit that there are causes in the universe which effect changes in the lower human soul, transforming it in the same way as do the causes, demonstrated by me, which lead the lower animal form into the higher one. If I do not admit this, the species of human souls remain for me a miracle in regard to their origin, just as the various animal species remain a miracle to the one who does not believe in the transformation of living creatures through laws of nature.

And this is absolutely correct: the modern believers, who deem themselves so greatly enlightened because they believe they have "cast out" the miracle in the domain of the living, are believers in miracles, nay, even worshipers of the miracle in the domain of the soul life. And only the following fact differentiates them from the believers in miracles, so greatly despised by them: these latter honestly avow their belief; the modern believers, however, have not the slightest inkling of the fact that they themselves have fallen prey to the darkest superstition.

And now let us illumine another corner of the "modern belief." In his Anthropology, Dr. Paul Topinard has beautifully compiled the findings of the modern theory of the origin of man. At the end of his book he briefly recapitulates the evolution of the higher animal forms in the various epochs of the earth according to Haeckel: "At the beginning of the earth period designated by geologists the Laurentian period, the first nuclei of albumin were formed by a chance meeting of certain elements, i.e. carbon, oxygen, hydrogen, and nitrogen, under conditions probably only prevailing at that epoch. From them, through spontaneous generation, monads developed (the smallest, imperfect living creatures). These split and multiplied, rearranged themselves into organs, and finally, after a series of transformations which Haeckel estimates as nine, they bestowed life upon certain vertebrae such as the amphioxus lanceolatus." We may skip the description of the further animal species in the same direction and add here at once Topinard's concluding sentences: "In the twentieth earth epoch, we find the anthropoid ape approximately during the whole Miocene period; in the twenty-first, the man-ape which does not yet possess speech and a corresponding brain. In the twenty-second period, Man finally appears as we know him, at least in his less perfect forms." And now, after having cited what is to be understood as the "natural-scientific basis of the modern belief," Topinard, in a few words, makes a significant confession. He says: "Here the classification comes to an abrupt halt. Haeckel forgets the twenty-third degree in which the brilliant Lamarck and Newton appear."

A corner in the creed of the modern believer is thereby exposed in which he points with the utmost clarity to facts, concerning which he denies his creed. He is unwilling to rise into the human soul sphere with the concepts with which he tried to find his way in the other spheres of nature. — Were he to do this, were he, with his attitude of mind acquired through the observation of external nature, to enter upon the sphere which Topinard calls the twenty-third degree, then he would have to say to himself: just as I derive the higher animal species from the lower through evolution, so do I derive the higher soul nature from the lower through evolution. I cannot understand Newton's soul if I do not conceive of it as having sprung from a preceding soul being. And this soul being can never be looked for in the physical ancestors. Were I to look for it there, I would turn upside down the whole method of nature research. How could it ever occur to a scientist to show the evolution of one animal species out of another if the latter, in regard to its physical makeup, were as dissimilar to the former as Newton, in regard to his soul, is to his forebears: One conceives of one animal species having proceeded from a similar one which is merely one degree lower than itself. Therefore, Newton's soul must have sprung from a soul similar to it, but only one degree lower, psychically. Newton's soul nature is comprised in his biography. I recognize Newton by his biography just as I recognize a lion by the description of its species. And I comprehend the species "lion" if I imagine that it has sprung from a species on a correspondingly lower stage. Thus I comprehend what is comprised in Newton's biography if I conceive of it as having developed from the biography of a soul which resembles it, is related to it as soul. From this follows that Newton's soul existed already in another form, just as the species "lion" existed previously in a different form.

For clear thought, there is no escape from this conception. Only because the modern believers do not have the courage to think their thoughts through to the end do they not arrive at this final conclusion. Through it, however, the reappearance of the being who is comprised in the biography is secured. — Either we must abandon the whole natural-scientific theory of evolution, or we must admit that it must be extended to include the evolution of the soul. There are only two alternatives: either, every soul is created by a miracle, just as the animal species would have to be created by miracles if they have not developed one out of the other, or, the soul has developed and has previously existed in another form, just as the animal species has existed in another form.

A few modern thinkers who have preserved some clarity and courage for logical thinking are a living proof of the above conclusion. They are just as unable to familiarize themselves with the thought of soul evolution, so strange to our age, as are the modern believers characterized above. But they at least possess the courage to confess the only other possible view, namely: the miracle of the creation of the soul. Thus, in the book on psychology by Professor Johannes Rehmke, one of the best thinkers of our time, we may read the following: "The idea of creation ... appears to us ... to be the only one suited to render comprehensible the mystery of the origin of the soul." Rehmke goes so far as to acknowledge the existence of a conscious Universal-Being who, "as the only condition for the origin of the soul, would have to be called the creator of the soul." Thus speaks a thinker who is unwilling to indulge in gentle spiritual slumber after having grasped the physical life processes, yet who is lacking the capacity of acknowledging the idea that each individual soul has evolved out of its previous form of existence. Rehmke has the courage to accept the miracle, since he is unable to have the courage to acknowledge the anthroposophical view of the reappearance of the soul, of reincarnation. Thinkers in whom the natural-scientific striving begins to be developed logically must of necessity arrive at this view. Thus, in the book, Neuchristentum und reale Religion (Neo-Christianity and Real Religion), by Julius Baumann, professor of philosophy at the University of Goettingen, we find the following (twenty-second) paragraph among the thirty-nine paragraphs of a Sketch of a Summary of Real-Scientific Religion: "Just as in inorganic nature the physical-chemical elements and forces do not disappear but only change their combinations, so is this also to be assumed, according to the real scientific method, in respect of the organic and organic-spiritual forces. The Human soul as formal unity, as connecting Ego, returns in new human bodies and is thus enabled to pass through all the stages of human evolution."

Whoever possesses the full courage for the natural-scientific avowal of faith of the present age must arrive at this conception. This, however, must not be misunderstood; we do not maintain that the more prominent thinkers among the modern believers are cowardly persons, in the ordinary sense of the word. It needed courage, indescribable courage to carry to victory the natural-scientific view in face of the resisting forces of the nineteenth century. But this courage must be distinguished from the higher one in regard to logical thinking. Yet just those nature researchers of the present age who desire to erect a world conception out of the findings of their domain are lacking such logical thinking. For, is it not a disgrace if we have to hear a sentence like the following, which was pronounced by the Breslau chemist Albert Ladenburg, in a lecture at a recent (1903) Conference of scientists: "Do we know anything about a substratum of the soul? I have no such knowledge." After having made this confession, this same man continues: "What is your opinion concerning immortality? I believe that in regard to this question, more than in regard to any other, the wish is father to the thought, for I do not know a single scientifically proven fact which might serve as the basis for the belief in immortality." What would the learned gentleman say if we were confronted by a speaker who said: "I know nothing about chemical facts. I therefore deny the chemical laws, for I know not a single scientifically proven fact which might serve as the basis for these laws." Certainly, the professor would reply: "What do we care about your ignorance of chemistry? First study chemistry, then do your talking!" Professor Ladenburg does not know anything about a substratum of the soul; he, therefore, should not bother the world with the findings of his ignorance.

Just as the nature researcher, in order to understand certain animal forms, studies the animal forms out of which these former have evolved, so the psychologist, rooted in natural science, must, in order to understand a certain soul form, study the soul form out of which the former has evolved. The skull form of higher animals is explained by scientists as having arisen out of the transformation of the lower animal skull. Therefore, everything belonging to a soul's biography ought to be explained by them through the biography of the soul out of which this soul concerned has evolved. The later conditions are the effects of former ones. That is to say, the later physical conditions are the effects of former physical conditions; likewise, the later soul conditions are the effects of former soul conditions. This is the content of the Law of Karma which says: all my talents and deeds in my present life do not exist separately as a miracle, but they are connected as effect with the previous forms of existence of my soul and as cause with future ones.

Those who, with open spiritual eyes, observe human life and do not know this comprehensive law, or do not wish to acknowledge it, are constantly confronted by riddles of life. Let us quote one example for many. It is contained in Maurice Maeterlinck's book Le Temple Enseveli (The Buried Temple). This is a book which speaks of these riddles, which appear to present-day thinkers in a distorted shape because they are not conversant with the great laws in spiritual life of cause and effect, of Karma. Those who have fallen prey to the limited dogmas of the modern believers have no organ for the perception of such riddles. Maeterlinck puts [forth] one of these questions: "If I plunge into the water in zero weather in order to save my fellow man, or if I fall into the water while trying to push him into it, the consequences of the cold I catch will be exactly the same in both cases, and no power in heaven or earth beside myself or the man (if he is able to do so) will increase my suffering because I have committed a crime, or will relieve my pain because I performed a virtuous deed." Certainly; the consequences in question here appear to an observation which limits itself to physical facts to be the same in both cases. But may this observation, without further research, be considered complete? Whoever asserts this holds, as a thinker, the same view point as a person who observes two boys being taught by two different teachers, and who observes nothing else in this activity but the fact that in both cases the teachers are occupied with the two boys for the same number of hours and carry on the same studies. If he were to enter more deeply upon the facts, he would perhaps observe a great difference between the two cases, and he would consider it comprehensible that one boy grows up to be an inefficient man, while the other boy becomes an excellent and capable human being. — And if the person who is willing to enter upon soul-spiritual connections were to observe the above consequences for the souls of the human beings in question, he would have to say to himself: what happens there cannot be considered as isolated facts. The consequences of a cold are soul experiences, and I must, if they are not to be deemed a miracle, view them as causes and effects in the soul life. The consequences for the person who saves a life will spring from causes different from those for the criminal; or they will, in the one or the other case, have different effects. And if I cannot find these causes and effects in the present life of the people concerned, if all conditions are alike for this present life, then I must look for the compensation in the past and the future life. Then I proceed exactly like the natural scientist in the field of external facts; he, too, explains the lack of eyes in animals living in dark caves by previous experiences, and he presupposes that present-day experiences will have their effects in future formations of races and species.

Only he has an inner right to speak of evolution in the domain of outer nature who acknowledges this evolution also in the sphere of soul and spirit. Now, it is clear that this acknowledgment, this extension of knowledge of nature beyond nature is more than mere cognition. For it transforms cognition into life; it does not merely enrich man's knowledge, it provides him with the strength for his life's journey. It shows him whence he comes and whither he goes. And it will show him this whence and whither beyond birth and death if he steadfastly follows the direction which this knowledge indicates. He knows that everything he does is a link in the stream which flows from eternity to eternity. The point of view from which he regulates his life becomes higher and higher. The man who has not attained to this state of mind appears as though enveloped in a dense fog, for he has no idea of his true being, of his origin and goal. He follows the impulses of his nature, without any insight into these impulses. He must confess that he might follow quite different impulses, were he to illuminate his path with the light of knowledge. Under the influence of such an attitude of soul, the sense of responsibility in regard to life grows constantly. If the human being does not develop this sense of responsibility in himself, he denies, in a higher sense, his humanness. Knowledge lacking the aim to ennoble the human being is merely the satisfying of a higher curiosity. To raise knowledge to the comprehension of the spiritual, in order that it may become the strength of the whole life, is, in a higher sense, duty. Thus it is the duty of every human being to seek the understanding for the Whence and Whither of the Soul.

Knowledge of the State
Between Death and a New Birth

The following thoughts are intended as aphoristic sketches of a domain of knowledge that, in the form in which is it characterised here, is almost entirely rejected by the culture of our time. The aphoristic form has been chosen in order to give some idea of the fundamental character of this field of knowledge, and to show — at least in one direction — the prospects for life which it opens up. The narrow frame of an essay requires one to refer the reader to the literature of the subject for further information. The author is fully aware that precisely this form of presentation may easily be felt as presumptuous by many who, from the well-founded habits of thought of the culture of the day, must find what is here brought forward directly opposed to all that is scientific. It may be said in answer to this that the author, in spite of his 'spiritual-scientific' orientation, believes that he can agree with every scientist in his high estimation of the spirit and significance of scientific thinking. Only it seems clear to him that one can fully accept Natural Science without being thereby compelled to reject an independent Spiritual Science of the kind described here. A consequence of this relation to Natural Science will, at all events, be to guard true Spiritual Science from that amateurishness which is noticeable in many quarters to-day, and which usually indulges the more presumptuously in phrases about the 'crude materialism of Natural Science' the less the speakers are able to judge of the earnestness, rigour and scientific soundness of Natural Knowledge.

The writer wished to make these introductory remarks because the brevity of the discussions in this article may possibly obscure from the reader his attitude towards these matters.

He who speaks to-day of investigating the spiritual world encounters the sceptical objections of those whose habits of thought have been moulded by the outlook of Natural Science. His attention will be drawn to the blessings which this outlook has brought for a healthy development of human life, by destroying the illusions of a learning which professed to follow purely spiritual modes of cognition. Now these sceptical objections can be quite intelligible to the spiritual investigator. Indeed it ought to be perfectly clear to him that any kind of spiritual investigation which finds itself in conflict with established ideas of Natural Science cannot rest on a sure foundation. A spiritual investigator with a feeling for, and an understanding of the earnestness of scientific procedure, and insight into the achievements of Natural Knowledge for human life, will not wish to join the ranks of those who, from the standpoint of their 'spiritual sight,' criticise lightly the limitations of scientists, and imagine their own standpoint so much the higher the more every kind of Natural Knowledge is lost for them in unfathomable depths.

Natural Science and Spiritual Science could live in harmony if the former could rid itself of the erroneous belief that true spiritual investigation necessarily requires we [human beings] to reject attested knowledge of sensible reality and of the soul-life bound up with this. In this erroneous belief lies the source of innumerable misunderstandings which Spiritual Science has to encounter. Those who believe they stand, in their outlook on life, on the 'firm ground of Natural Science' hold that the spiritual investigator is compelled by his point of view to reject their knowledge. But this is not really the case. Genuine spiritual investigation is in full agreement with Natural Science. Thus spiritual investigation is not opposed on account of what it maintains, but for what people believe it could or must maintain.

With regard to human soul life the scientific thinker must maintain that the soul activities which reveal themselves as thinking, feeling and willing, ought, for the acquisition of scientific knowledge, to be observed without prejudice in the same way as the phenomena of light or heat in the outer world of Nature. He must reject all ideas about the entity of the soul which do not arise from such unprejudiced observation, and from which all kinds of conclusions are then drawn about the indestructibility of the soul, and its connection with the spiritual world. It is quite understandable that such a thinker begins his study of the facts of soul-life as Theodor Ziehen does in the first of his lectures on "Physiological Psychology." He says: "The psychology which I shall put before you, is not that old psychology which attempted to investigate soul phenomena in a more or less speculative way. This psychology has long been abandoned by those accustomed to think scientifically." True spiritual investigation need not conflict with the scientific attitude which may be in such an avowal. And yet, among those who take this attitude as a result of their scientific habits of thought, the opinion will be almost universally held to-day that the specific results of spiritual investigation are to be regarded as unscientific. Of course one will not encounter everywhere this rejection, on grounds of principle, of the investigation of spiritual facts; yet when specific results of such investigation are brought forward they will scarcely escape the objection that scientific thinking can do nothing with them. As a consequence of this, one can observe that there has recently grown up a science of the soul, forming its methods of investigation on the pattern of natural-scientific procedure, but unable to find the power to approach those highest questions which our inner need of knowledge must put when we turn our gaze to the fate of the soul. One investigates conscientiously the connection of soul phenomena with bodily processes, one tries to gain ideas on the way presentations associate and dissociate in the soul, how attention acts, how memory functions, what relation exists between thinking, feeling and willing; but for the higher questions of soul-life the words of Franz Brentano remain true. This acute psychologist, though rooted in the mode of thinking of Natural Science, wrote: "The laws of association of ideas, of the development of convictions and opinions and of the genesis of pleasure and love would be anything but a true compensation for the hopes of a Plato or an Aristotle of gaining certainty concerning the continued life of our better part after the dissolution of the body." And if the recent scientific mode of thinking really means "excluding the question of immortality," this exclusion would have great significance for psychology.

The fact is, that considerations which might tend in the direction of the 'hopes of a Plato and an Aristotle' are avoided in recent psychological writings which wish to satisfy the demands of scientific thought. Now the spiritual investigator will not come into conflict with the mode of procedure of recent scientific psychology if he has an understanding of its vital nerve. He will have to admit that this psychology proceeds, in the main, along right lines insofar as the study of the inner experiences of thinking, feeling and willing is concerned. Indeed his path of knowledge leads him to admit that thinking, feeling and willing reveal nothing that could fulfil the 'hopes of a Plato and an Aristotle' if these activities are only studied as they are experienced in ordinary human life. But his path of knowledge also shows that in thinking, feeling and willing something lies hidden which does not become conscious in the course of ordinary life, but which can be brought to consciousness through inner soul exercises. In this spiritual entity of the soul, hidden from ordinary consciousness, is revealed what in it is independent of the life of the body; and in this the relations of man to the spiritual world can be studied. To the spiritual investigator it appears just as impossible to fulfil the 'hopes of a Plato or an Aristotle' in regard to the existence of the soul independent of bodily life by observing ordinary thinking, feeling and willing, as it is impossible to investigate in water the properties of hydrogen. To learn these one must first extract the hydrogen from the water by an appropriate procedure. So it is also necessary to separate from the everyday life of the soul (which it leads in connection with the body) that entity which is rooted in the spiritual world, if this entity is to be studied.

The error which casts befogging misunderstandings in the way of Spiritual Science lies in the almost general belief that knowledge about the higher questions of soul-life must be gained from a study of such facts of the soul as are already to be found in ordinary life. But no other knowledge results from these facts than that to which research, conducted on what are at present called scientific lines, can lead. On this account Spiritual Science can be no mere heeding of what is immediately present in the life of the soul. It must first lay bare, by inner processes in the life of the soul, the world of facts to be studied. To this end spiritual investigation applies soul processes which are attained in inner experience. Its field of research lies entirely within the inner life of the soul. It cannot make its experiences outwardly visible. Nevertheless they are not on that account less independent of personal caprice than the true results of Natural Science. They have nothing in common with mathematical truths except that they, too, cannot be proved by outer facts, but are proved for anyone who grasps them in inner perception. Like mathematical truths they can at the most be outwardly symbolised but not represented in their full content, for it is this that proves them. The essential point, which can easily be misunderstood, is, that on the path pursued by spiritual investigation a certain direction is given, by inner initiative, to the experiences of the soul, thereby calling out forces which otherwise remain unconscious as in a kind of soul sleep. (The soul exercises which lead to this goal are described in detail in my books "Knowledge of the Higher Worlds and its Attainment" and "Occult Science." It is only intended to indicate here what transpires in the soul when it subjects itself to such exercises). If the soul proceeds in this way it inserts — as it were — its inner life into the domain of spiritual reality. It opens to the spiritual world its organs of perception so formed, as the senses open outwardly to physical reality.

One kind of such soul exercises consists in an intensive surrender to the process of thinking. One carries this surrender so far that one acquires the capacity of directing one's attention no longer to the thoughts present in thinking but solely to the activity of thinking itself. Every kind of thought content then disappears from consciousness and the soul experiences herself consciously in the activity of thinking. Thinking then becomes transformed into a subtle inner act of will which is completely illuminated by consciousness. In ordinary thinking, thoughts live; the process indicated extinguishes the thought in thinking. The experience thus induced is a weaving in an inner activity of will which bears its reality within itself. The point is that the soul, by continued inner experience in this direction, may make itself as familiar with the purely spiritual reality in which it weaves as sense observation is with physical reality. As in the outer world a reality can only be known as such by experiencing it, so, too, in this inner domain. He who objects that what is inwardly real cannot be proved only shows that he has not yet grasped that we become convinced of an outer reality in no other way than by experiencing its existence together with our own. A healthy life has direct experience of the difference between a genuine perception in the outer world and a vision or hallucination; in a similar way a healthily developed soul life can distinguish the spiritual reality it has approached from fantastic imagining; and dreamy reverie.

Thinking that has been developed in the manner stated perceives that it has freed itself from the soul force which ordinarily leads to memory. What is experienced in thinking which has become an inwardly experienced 'will-reality' cannot be remembered in the direct form in which it presents itself. Thus it differs from what is experienced in ordinary thinking. What one has thought about an event is incorporated into memory. It can be brought up again in the further course of life. But the 'will-reality' here described must be attained anew, if it is to be again experienced in consciousness. I do not mean that this reality cannot be indirectly incorporated into ordinary memory. This must indeed take place if the path of spiritual investigation is to be a healthy one. But what remains in memory is only an idea (Vorstellung) of this reality, just as what one remembers to-day of an experience of yesterday is only an idea (Vorstellung). Concepts, ideas, can be retained in memory: a spiritual reality must be experienced ever anew. By grasping vividly this difference between the cherishing of mere thoughts and a spiritual reality reached by developing the activity of thinking, one comes to experience oneself with this reality outside the physical body. What ordinary thinking must mostly regard as an impossibility commences; one experiences oneself outside the existence that is connected with the body. Ordinary thinking, regarding this experience 'outside the body' only from its own point of view, must at first hold this to be an illusion. Assurance of this experience can, indeed, only be won through the experience itself. And it is precisely through this experience that one understands only too well that those whose habits of thought have been formed by Natural Science cannot, at first, but regard such experiences as fantastic imaginings or dreamy reverie, perhaps as a weaving in illusions or hallucinations. Only he can fully understand what is here brought forward who has come to know that the path of true spiritual investigation releases forces in the soul which lie in a direction precisely opposite to those which induce pathological soul experiences. What the soul develops on the path of spiritual investigation are forces competent to oppose pathological states or to dissipate these where they tend to occur. No scientific investigation can see through what is visionary — of an hallucinatory nature — when this tries to get in man's way, as directly as true spiritual science, which can only unfold in a direction opposed to the unhealthy experiences mentioned.

In that moment when this 'experience outside the body' becomes a reality for him the spiritual investigator learns to know how ordinary thinking is bound to the physical processes of the body. He comes to see how thoughts acquired in outer experience necessarily arise in such a way that they can be remembered. This rests on the fact that these thoughts do not merely lead a spiritual life in the soul but share their life with the body. Thus the spiritual investigator comes not to reject but to accept what scientific thought must maintain about the dependence of the life of thought on bodily processes.

At first the inner experiences described above present themselves as anxious oppression of the soul. They appear to lead out of the domain of ordinary existence but not into a new reality. One knows, indeed, that one is living in a reality; one feels this reality as one's own spiritual being. One has found one's way out of sense reality, but one has only grasped oneself in a purely spiritual form of existence. A feeling of loneliness resembling fear can overtake the soul — a loneliness to experience oneself in a world, not merely to possess oneself. Yet another feeling arises. One feels one must lose again the acquired spiritual self-experience, if one cannot confront a spiritual environment. The spiritual state into which one thus enters may be roughly compared to what would be experienced if one had to clutch with one's hands in all directions without being able to lay hold of anything.

When, however, the path of spiritual investigation is pursued in the right way, the above experiences are, indeed, undergone, but they form only one side of the soul's development. The necessary completion is found in other experiences. As certain impulses given to the soul's experiences lead one to grasp the 'will-reality' within thinking, so other directions imparted to the processes of the soul lead to an experience of hidden forces within the activity of the will. (Here also we can only state what takes place in the inner being of man through such soul experiences. The books mentioned give a detailed description of what the soul must undertake in order to reach the indicated goal). In ordinary life the activity of the will is not perceived in the same way as an outer event. Even what is usually called introspection by no means puts one into the position of regarding one's own willing as one regards an outer event of Nature.

To achieve this — to be able to confront one's own willing as an observer stands before an outer fact of Nature — intensive soul processes, induced voluntarily, are again necessary. If these are induced in the appropriate way there arises something quite different from this view of one's own willing as of an outer fact. In ordinary perception a presentation (Vorstellung) emerges in the life of the soul and is, in a certain sense, an inner image of the outer fact. But in observing one's own willing this accustomed power of forming presentations fades out. One ceases to form presentations of outer things. In place of this a faculty of forming real images — a real perception — is released from the depths of willing, and breaks through the surface of the will's activity, bringing living spiritual reality with it. At first one's own hidden spiritual entity appears within this spiritual reality. One perceives that one carries a hidden spiritual man within one. This is no thought-picture but a real being — real in a higher sense than the outer bodily man. Now this spiritual man does not present himself like an outer being perceptible to the senses. He does not reveal his characteristic qualities outwardly. He reveals himself through his inner nature by developing an inner activity similar to the processes of consciousness in one's own soul. But, unlike the soul dwelling in man's body, this higher being is not turned towards sensible objects but towards spiritual events — in the first place towards the events of one's own soul-life as unfolded up till now. One really discovers in oneself a second human being who, as a spiritual being, is a conscious observer of one's ordinary soul-life. However fantastic this description of a spiritual man within the bodily may appear, it is nevertheless a sober description of reality for a soul-life appropriately trained. It is as different from anything visionary or of the nature of an illusion as is day from night.

Just as a reality partaking of the nature of will is discovered in the transformed thinking, so a consciousness partaking of the nature of being — and weaving in the spiritual — is discovered in the will. And these two prove, for fuller experience, to belong together. In a certain sense they are discovered on paths running in opposite directions, but turn out to be a unity. The feeling of anxiety experienced in the weaving of the 'will-reality' ceases when this 'will-reality,' born from developed thinking, unites itself with the higher being above described. Through this union man confronts, for the first time, the complete spiritual world. He encounters, not only himself, but beings and events of the spiritual world lying outside himself.

In the world into which man has thus entered, perception is an essentially different process from perception in the world of sense. Real beings and events of the spiritual world arise from out of the higher being revealed through developing the will. Through the interplay of these beings and events with the 'will-reality' resulting from developed thinking, these beings and events are spiritually perceived. What we know as memory in the physical world ceases to have significance for the spiritual world. We see that this soul force uses the physical body as a tool. But another force takes the place of memory in observing the spiritual world. Through this force a past event is not remembered in the form of mental presentations but perceived directly in a fresh experience. It is not like reading a sentence and remembering it later, but like reading and re-reading. The concept of the past acquires a new significance in this domain; the past appears to spiritual perception as present, and we recognise that something belongs to a past time by perceiving, not the passage of time, but the relation of one spiritual being or event to another.

The path into the spiritual world is thus traversed by laying bare what is contained in thinking and willing. Now feeling cannot be developed in a similar way by inner initiative of soul. Unlike the case of thinking and willing, nothing to take the place of what is experienced within the physical world as feeling can be developed in the spiritual world through transforming an inner force. What corresponds to feeling in the spiritual world arises quite of itself as soon as spiritual perception has been acquired in the described way. This experience of feeling, however, bears a different character from that borne by feeling in the physical world. One does not feel in oneself, but in the beings and events which one perceives. One enters into them with one's feeling; one feels their inner being, as in physical life one feels one's own being. We might put it in this way: as in the physical world one is conscious of experiencing objects and events as material, so in the spiritual world one is conscious of experiencing beings and facts through revelations of feeling which come from without like colours or sounds in the physical world.

A soul which has attained to the spiritual experience described knows it is in a world from out of which it can observe its own experiences in the physical word — just as physical perception can observe a sensible object. It is united with that spiritual entity which unites itself — at birth (or at conception) — with the physical body derived from one's ancestors; and this spiritual entity persists when this body is laid aside at death. The 'hopes of a Plato and an Aristotle' for the science of the soul can only be fulfilled through a perception of this entity. Moreover the perception of repeated earth-lives (between which are lives spent in the purely spiritual world) now becomes a fact inasmuch as man's psychic-spiritual kernel, thus discovered, perceives itself and its own weaving and becoming in the spiritual world. It learns to know its own being as the result of earlier earthlives and spiritual forms of existence lying between them. Within its present earth-life it finds a spiritual germ which must unfold in a future earth-life after passing through states between death and a new birth. As the plant germ contains the future plant potentially, so there develops, concealed in man, a psychic-spiritual germ. This reveals itself to spiritual perception through its own essence as the foundation of a future earth-life. It would be incorrect so to interpret the spiritual perception of life between death and a new birth as if such perception meant participating beforehand in the experience of the spiritual world entered at physical death. Such perception does not give a complete, disembodied experience of the spiritual world as experienced after death; it is only the knowledge of the actual experience that is experienced.

While still in one's body one can receive all of the disembodied experience between death and a new birth that is offered by the experiences of the soul described above, that is to say, when the 'will-reality' is released from thinking with the help of the consciousness set free from the will. In the spiritual world the feeling element revealing itself from without can first be experienced through entrance into this world. Strange as it may sound, experience in the spiritual world leads one to say: the physical world is present to man in the first place as a complex of outer facts, and man acquires knowledge of it after it has confronted him in this form; the spiritual world, on the other hand, sends knowledge of itself in advance, and the knowledge it kindles in the soul beforehand is the torch which must illumine the spiritual world if this world is to reveal itself as a fact. It is clear to one who knows this through spiritual perception that this light develops during bodily life on earth in the unconscious depths of the soul, and then, after death, illumines the regions of the spiritual world making them experiences of the human soul.

During bodily life on earth one can awaken this knowledge of the state between death and a new birth. This knowledge has an entirely opposite character to that developed for life in the physical world. One perceives through it what the soul will accomplish between death and a new birth, because one has present in spiritual perception the germ of what impels towards this accomplishment. The perception of this germ reveals that a creative connection with the spiritual world commences for the soul after death. It unfolds an activity which is directed towards the future earth life as its goal, whereas in physical perception its activity is directed — although imitatively and not creatively — towards the outer world of sense. Man's growth (Werden) as a spiritual being connected with the spiritual world lies in the field of vision of the soul between death and a new birth, as the existence (Sein) of the sense world lies in the field of view of the bodily man. Active perception of spiritual Becoming (Werden) characterises the conditions between death and a new birth. (It is not the task of this article to give details of these states. Those interested will find them in my books "Theosophy" and "Occult Science").

In contrast to experience in the body, spiritual experience is something to which we are completely unaccustomed, inasmuch as the idea of Being as acquired in the physical world loses all meaning. The spiritual world has nothing of the nature of Being. Everything is Becoming. To enter a spiritual environment is to enter an everlasting Becoming. But in contrast to this restless Becoming in our spiritual environment we have the soul's perception of itself as stationary consciousness within the never-ceasing movement into which it is placed. The awakened spiritual consciousness must accommodate itself to this reversal of inner experience with regard to the consciousness that lives in the body. It can thereby acquire a real knowledge of experience apart from the body. And only such knowledge can embrace the states between death and a new birth.

" . . . . In a certain sense all human beings are 'specialists' to-day so far as their souls are concerned. We are struck by this specialised mode of perception when we study the development of Art in humanity. And for this very reason a comprehensive understanding of spiritual life in its totality must again come into existence. True form in Art will arise from this comprehensive understanding of spiritual life . . . . "

Supersensible Knowledge

There are two experiences whence the soul may gain an understanding for the mode of knowledge to which the supersensible worlds will open out. The one originates in the science of Nature; the other, in the Mystical experience whereby the untrained ordinary consciousness contrives to penetrate into the supersensible domain. Both confront the soul of man with barriers of knowledge — barriers he cannot cross till he can open for himself the portals which by their very essence Natural Science, and ordinary Mysticism too, must hold fast closed.

Natural Science leads inevitably to certain conceptions about reality, which are like a stone wall to the deeper forces of the soul; and yet, this Science itself is powerless to remove them. He who fails to feel the impact, has not yet called to life the deeper needs of knowledge in his soul. He may then come to believe that it is impossible in any case for Man to attain any other than the natural-scientific form of knowledge. There is, however, a definite experience in Self-knowledge whereby one weans oneself of this belief. This experience consists in the insight that the whole of Natural Science would be dissolved into thin air if we attempted to fathom the above-named conceptions with the methods of Natural Science itself. If the conceptions of Natural Science are to remain spread out before the soul, these limiting conceptions must be left within the field of consciousness intact, without attempting to approach them with a deeper insight. There are many of them; here I will only mention two of the most familiar: Matter and Force. Recent developments in scientific theory may or may not be replacing these particular conceptions; the fact remains that Natural Science must invariably lead to some conception or another of this kind, impenetrable to its own methods of knowledge.

To the experience of soul, of which I am here speaking, these limiting conceptions appear like a reflecting surface which the human soul must place before it; while Natural Science itself is like the picture, made manifest with the mirror's help. Any attempt to treat the limiting conceptions themselves by ordinary scientific means is, as it were, to smash the mirror, and with the mirror broken, Natural Science itself dissolves away. Moreover, this experience reveals the emptiness of all talk about 'Things-in-themselves,' of whatsoever kind, behind the phenomena of Nature. He who seeks for such

Things-in-themselves is like a man who longs to break the looking-glass, hoping to see what there is behind the reflecting surface to cause his image to appear.

It goes without saying that the validity of such an experience of soul cannot be 'proved,' in the ordinary sense of the word, with the habitual thoughts of presentday Natural Science. For the point will be, what kind of an inner experience does the process of the 'proof' call forth in us; and this must needs transcend the abstract proof. With inner experience in this sense, we must apprehend the question: How is it that the soul is forced to confront these barriers of knowledge in order to have before it the phenomena of Nature? Mature self-knowledge brings us an answer to this question. We then perceive which of the forces of man's soul partakes in the erection of these barriers to knowledge. It is none other than the force of soul which makes man capable, within the world of sense, of unfolding Love out of his inner being. The faculty of Love is somehow rooted in the human organisation; and the very thing which gives to man the power of love — of sympathy and antipathy with his environment of sense, — takes away from his cognition of the things and processes of Nature the possibility to make transparent such pillars of Reality as 'Matter' and 'Force.' To the man who can experience himself in true self-knowledge, on the one hand in the act of knowing Nature, and on the other hand in the unfolding of Love, this peculiar property of the human organisation becomes straightway apparent.

We must, however, beware of misinterpreting this perception by lapsing again into a way of thought which, within Natural Science itself, is no doubt inevitable. Thus it would be a misconstruction to assume, that an insight into the true essence of the things and processes of Nature is withheld from man because he lacks the organisation for such insight. The opposite is the case. Nature becomes sense-perceptible to man through the very fact that his being is capable of Love. For a being incapable of Love within the field of sense, the whole human picture of Nature would dissolve away. It is not Nature who on account of his organisation reveals only her external aspect. No; it is man, who, by that force of his organisation which makes him in another direction capable of Love, is placed in a position to erect before his soul images and forms of Reality whereby Nature reveals herself to him.

Through the experience above-described the fact emerges, that the scientific frontiers of knowledge depend on the whole way in which man, as a sense-endowed being, is placed within this world of physical reality. His vision of Nature is of a kind, appropriate to a being who is capable of Love. He would have to tear the faculty of Love out of his inner life if he wished no longer to be faced with limits in his perception of Nature. But in so doing he would destroy the very force whereby Nature is made manifest to him. The real object of his quest for knowledge is not, by the same methods which he applies in his outlook upon Nature, to remove the limitations of that outlook. No, it is something altogether different, and once this has been perceived, man will no longer try to penetrate into a supersensible world through the kind of knowledge which is effective in Natural Science. Rather will he tell himself, that to unveil the supersensible domain an altogether different activity of knowledge must be evolved than that which he applies to the science of Nature.

Many people, more or less consciously aware of the above experience of soul, turn away from Natural Science when it is a question of opening the supersensible domain, and seek to penetrate into the latter by methods which are commonly called Mystical. They think that what is veiled to outwardly directed vision may be revealed by plunging into the depths of one's own being. But a mature self-knowledge reveals in the inner life as well a frontier of knowledge. In the field of the senses the faculty of Love erects, as it were, an impenetrable background whereat Nature is reflected; in the inner life of man the power of Memory erects a like background. The same force of soul, which makes the human being capable of Memory, prevents his penetrating, in his inner being, down to that experience which would enable him to meet — along this inward path — the supersensible reality for which he seeks. Invariably, along this path, he reaches only to that force of soul which recalls to him in Memory the experiences he has undergone through his bodily nature in the past. He never penetrates into the region where with his own supersensible being he is rooted in a supersensible world. For those who fail to see this, mystical pursuits will give rise to the worst of illusions. For in the course of life, the human being receives into his inner life untold experiences, of which in the receiving he is not fully conscious. But the Memory retains what is thus half-consciously or subconsciously experienced. Long afterwards it frequently emerges into consciousness — in moods, in shades of feeling and the like, if not in clear conceptions. Nay more, it often undergoes a change, and comes to consciousness in quite a different form from that in which it was experienced originally. A man may then believe himself confronted by a supersensible reality arising from the inner being of the soul, whereas, in fact, it is but an outer experience transformed — an experience called forth originally by the world of sense — which comes before his mental vision. He alone is preserved from such illusions, who recognises that even on a mystic path man cannot penetrate into the supersensible domain so long as he applies methods of knowledge dependent on the bodily nature which is rooted in the world of sense. Even as our picture of Nature depends for its existence on the faculty of Love, so does the immediate consciousness of the human Self depend upon the power of Memory. The same force of the soul, endowing man in the physical world with the Self-consciousness that is bound to the bodily nature, stands in the way to obstruct his inner union with the supersensible world. Thus, even that which is often considered Mysticism provides no way into the supersensible realms of existence.

For him who would penetrate with full conscious clarity of understanding into the supersensible domain, the two experiences above described are, however, preparatory stages. Through them he recognises that man is shut off from the supersensible world by the very thing which places him, as a self-conscious being, in the midst of Nature. Now one might easily conclude from this, that man must altogether forego the effort to gain knowledge of the Supersensible. Nor can it be denied that many who are loath to face the painful issue, abstain from working their way through to a clear perception of the two experiences. Cherishing a certain dimness of perception on these matters, they either give themselves up to the belief that the limitations of Natural Science may be transcended by some intellectual and philosophic exercise; or else they devote themselves to Mysticism in the ordinary sense, avoiding the full enlightenment as to the nature of Self-consciousness and Memory which would reveal its insufficiency.

But to one who has undergone them and reached a certain clarity withal, these very experiences will open out the possibility and prospect of true supersensible knowledge. For in the course of them he finds that even in the ordinary action of human consciousness there are forces holding sway within the soul, which are not bound to the physical organisation; forces which are in no way subject to the conditions whereon the faculties of Love and Memory within this physical organisation depend. One of these forces reveals itself in Thought. True, it remains unnoticed in the ordinary conscious life; indeed there are even many philosophers who deny it. But the denial is due to an imperfect self-observation. There is something at work in Thought which does not come into it from the faculty of Memory. It is something that vouches to us for the correctness of a present thought, not when a former thought emerging from the memory sustains it, but when the correctness of the present thought is experienced directly. This experience escapes the every-day consciousness, because man completely spends the force in question for his life of thought-filled perception. In Perception permeated by Thought this force is at work. But man, perceiving, imagines that the perception alone is vouching for the correctness of what he apprehends by an activity of soul where Thought and Perception in reality always flow together. And when he lives in Thought alone, abstracted from perceptions, it is but an activity of Thought which finds its supports in Memory. In this abstracted Thought the physical organism is cooperative. For the every-day consciousness, an activity of Thought unsubjected to the bodily organism is only present while man is in the act of Sense-perception. Sense-perception itself depends upon the organism. But the thinking activity, contained in and co-operating with it, is a purely supersensible element in which the bodily organism has no share. In it the human soul rises out of the bodily organism. As soon as man becomes distinctly, separately conscious of this Thinking in the act of Perception, he knows by direct experience that he has himself as a living soul, quite independently of the bodily nature.

This is man's first experience of himself as a supersensible soul-being, arising out of an evolved self-knowledge. The same experience is there unconsciously in every act of perception. We need only sharpen our selfobservation so as to Observe the fact: in the act of Perception a supersensible element reveals itself. Once it is thus revealed, this first, faint suggestion of an experience of the soul within the Supersensible can be evolved, as follows: In living, meditative practice, man unfolds a Thinking wherein two activities of the soul flow together, namely that which lives in the ordinary consciousness in Sense-perception, and that which is active in ordinary Thought. The meditative life thus becomes an intensified activity of Thought, receiving into itself the force that is otherwise spent in Perception. Our Thinking in itself must grow so strong, that it works with the same vivid quality which is otherwise only there in Sense-perception. Without perception by the senses we must call to life a Thinking which, unsupported by memories of the past, experiences in the immediate present a content of its own, such as we otherwise only can derive from Sense-perception. From the Thinking that co-operates in perception, this meditative action of the soul derives its free and conscious quality, its inherent certainty that it receives no visionary content raying into the soul from unconscious organic regions. A visionary life of whatsoever kind is the very antithesis of what is here intended. By self-observation we must become thoroughly and clearly familiar with the condition of soul in which we are in the act of perception through any one of the senses. In this state of soul, fully aware that the content of our ideation does not arise out of the activity of the bodily organism, we must learn to experience ideas which are called forth in consciousness without external perceptions, just as are those of which we are conscious in ordinary life when engaged in reflective thought, abstracted from the enter world. (As to the right ways of developing this meditative practice, detailed indications are given in the book 'Knowledge of the Higher Worlds and its Attainment' and in several of my other writings.)

In evolving the meditative life above-described, the human soul rises to the conscious feeling perception of itself, as of a supersensible Being independent of the bodily organisation. This is man's first experience of himself as a supersensible Being; and it leads on to a second stage in supersensible self-knowledge. At the former stage he can only be aware that he is a supersensible Being; at the second he feels this Being filled with real content, even as the 'I' of ordinary waking life is felt by means of the bodily organisation. It is of the utmost importance to realise that the transition from the one stage to the other takes place quite independently of any co-operation from outside the soul's domain — namely from the mere organic life. If we experienced the transition, in relation to our own bodily nature, any differently from the process of drawing a logical conclusion for example, it would be a visionary experience, not what is intended here. The process here intended differs from the act of drawing logical conclusions, not in respect of its relationship to the bodily nature, but in quite another regard; namely in the consciousness that a supersensible, purely spiritual content is entering the feeling and perception of the Self.

The kind of meditative life hitherto described gives rise to the supersensible self-consciousness. But this self-consciousness would be left without any supersensible environment if the above form of meditation were unaccompanied by another. We come to an understanding of this latter kind by turning our self-observation to the activity of the Will. In every-day life the activity of the Will is consciously directed to external actions. There is, however, another concomitant expression of the Will to which the human being pays little conscious attention. It is the activity of Will which carries him from one stage of development to another in the course of life. For not only is he filled with different contents of soul day after day; his soul-life itself, on each succeeding day, has evolved out of his soul-life of the day before. The driving force in this evolving process is the Will, which in this field of its activity remains for the most part unconscious. Mature self-knowledge can, however, raise this Will, with all its peculiar quality, into the conscious life. When this is done, man comes to the perception of a life of Will which has absolutely nothing to do with any processes of a sense-perceptible external world, but is directed solely to the inner evolution of the soul — independent of this world. Once it is known to him, he learns by degrees to enter into the living essence of this Will, just as in the former kind of meditative life he entered into the fusion of the soul's experiences of Thinking and Perception. And the conscious experience in this element of Will expands into the experience of a supersensible external world. Evolved in the way above described, and transplanted now into this element of Will, the supersensible self-consciousness finds itself in a supersensible environment, filled with spiritual Beings and events. While the supersensible Thinking leads to a self-consciousness independent of the power of Memory which is bound to the bodily nature, the supersensible Willing comes to life in such a way as to be permeated through and through by a spiritualised faculty of Love. It is this faculty of Love which enables the supersensible self-consciousness of man to perceive and grasp the supersensible external world. Thus the power of supersensible knowledge is established by a self-consciousness which eliminates the ordinary Memory and lives in the intuitive perception of the spiritual world through the power of Love made spiritual.

Only by realising this essence of the supersensible faculty of knowledge, does one become able to understand the real meaning of man's knowledge of Nature. In effect, the knowledge of Nature is inherently connected with what is being evolved in man within this physical world of sense. It is in this world that man incorporates, into his spiritual Being, Self-consciousness and the faculty of Love. Once he has instilled these two into his nature, he can carry them with him into the super sensible world. In supersensible perception, the ordinary power of Memory is eliminated. Its place is taken by an immediate vision of the past — a vision for which the past appears as we look backward in spiritual observation, just as for sense-perception the things we pass by as we walk along appear when we turn round to look behind us. Again the ordinary faculty of Love is bound to the physical organism. In conscious supersensible experience, its place is taken by a power of Love made spiritual, which is to say, a power of perception.

It may already be seen, from the above description, that supersensible experience takes place in a mood of soul which must be held apart, in consciousness, from that of ordinary Perception, Thinking, Feeling and Willing. The two ways of looking out upon the world must be kept apart by the deliberate control of man himself, just as in another sphere the waking consciousness is kept apart from the dream life. He who lets play the picture-complexes of his dreams into his waking life becomes a listless and fantastic fellow, abstracted from realities. He, on the other hand, who holds to the belief that the essence of causal relationships experienced in waking life can be extended into the life of dreams, endows the dream-pictures with an imagined reality which will make it impossible for him to experience their real nature. So with the mode of thought which governs our outlook upon Nature, or of inner experience which determines ordinary Mysticism: — he who lets them play into his supersensible experience, will not behold the supersensible, but weave himself in figments of the mind, which, far from bringing him nearer to it, will cut him off from the higher world he seeks. A man who will not hold his experience in the supersensible apart from his experience in the world of the physical senses, will mar the fresh and unembarrassed outlook upon Nature which is the true basis for a healthy sojourn in this earthly life. Moreover, he will permeate with the force of spiritual perception the faculty of Love that is connected with the bodily nature, thus tending to bring it into a deceptive relationship with the physical experience. All that the human being experiences and achieves within the field of sense, receives its true illumination — an illumination which the deepest needs of the soul require — through the science of things that are only to be experienced supersensibly. Yet must the latter be held separate in consciousness from the experience in the world of sense. It must illumine our knowledge of Nature, our ethical and social life; yet so, that the illumination always proceeds from a sphere of experience apart. Mediately, through the attunement of the human soul, the Supersensible must indeed shed its light upon the Sensible. For if it did not do so, the latter would be relegated to darkness of thought, chaotic wilfulness of instinct and desire.

Many human beings, well knowing this relationship which has to be maintained in the soul between the experience of the supersensible and that of the world of sense, hold that the supersensible knowledge must on no account be given full publicity. It should remain, so they consider, the secret knowledge of a few, who have attained by strict self-discipline the power to establish and maintain the true relationship. Such guardians of supersensible knowledge base their opinion on the very true assertion that a man who is in any way inadequately prepared for the higher knowledge will feel an irresistible impulse to mingle the Supersensible with the Sensible in life; and that he will inevitably thus call forth, both in himself and others, all the ill effects which we have here characterised as the result of such confusion. On the other hand — believing as they do, and with good reason, that man's outlook upon Nature must not be left to grope in utter darkness, nor his life to spend itself in blind forces of instinct and desire, — they have founded self-contained and closed Societies, or Occult Schools, within which human beings properly prepared are guided stage by stage to supersensible discovery.

Of such it then becomes the task to pour the fruits of their knowledge into life, without, however, exposing the knowledge itself to publicity.

In past epochs of human evolution this idea was undoubtedly justified. For the propensity above described, leading to the misuse of supersensible knowledge, was then the only thing to be considered, and against it there stood no other circumstance to call for publication of the higher knowledge. It might at most be contended that the superiority of those initiated into the higher knowledge gave into their hands a mighty power to rule over those who had no such knowledge.

None the less, an enlightened reading of the course of History will convince us that such conflux of power into the hands of a few, fitted by self-discipline to wield it, was indeed necessary.

In present time, however — meaning 'present' in the wider sense — the evolution of mankind has reached a point whenceforward it becomes not only impossible but harmful to prolong the former custom. The irresistible impulsion to misuse the higher knowledge is now opposed by other factors, making the — at any rate partial — publication of such knowledge a matter of necessity, and calculated also to remove the ill effects of the above tendency. Our knowledge of Nature has assumed a form wherein it beats perpetually, in a destructive way, against its own barriers and limitations. In many branches of Science, the laws and generalisations in which man finds himself obliged to clothe certain of the facts of Nature, are in themselves of such a kind as to call his attention to his own supersensible powers. The latter press forward into the conscious life of the soul. In former ages, the knowledge of Nature which was generally accessible had no such effect. Through Natural Science, however, in its present form — expanding as it is in ever widening circles — mankind would be led astray in either of two directions, if a publication of supersensible knowledge were not now to take place. Either the possibility of a supersensible world-outlook would be repudiated altogether and with growing vehemence; and this would presently result in an artificial repression of supersensible faculties which the time is actually calling forth. Such repression would make it more and more impossible for man to see his own Being in a true light. Emptiness, chaos and dissatisfaction of the inner life, instability of soul, perversity of will; and, in the sequel, even physical degeneration and illhealth would be the outcome. Or else the supersensible faculties-uncontrolled by conscious knowledge of these things-would break out in a wild tangle of obtuse, unconscious, undirected forces of cognition, and the life of knowledge would degenerate in a chaotic mass of nebulous conceptions. This would be to create a world of scientific phantoms, which, like a curtain, would obscure the true supersensible world from the spiritual eye of man. For either of these aberrations, a proper publication of supersensible knowledge is the only remedy.

As to the impulse to abuse such knowledge in the way above described, it can be counteracted in our time, as follows: the training of thought which modern Natural Science has involved can be fruitfully employed to clothe in words the truths that point towards the supersensible. Itself, this Science of Nature cannot penetrate into the supersensible world; but it lends the human mind an aptitude for combinations of thought whereby the higher knowledge can be so expressed that the irresistible impulsion to misuse it need not arise. The thought-combinations of the Nature-knowledge of former times were more pictorial, less inclined to the domain of pure Thought. Supersensible perceptions, clothed in them, stirred up — without his being conscious of it — those very instincts in the human being which tend towards misuse.

This being said, it cannot on the other hand be emphasised too strongly that he who gives out supersensible knowledge in our time will the better fulfil his responsibilities to mankind the more he contrives to express this knowledge in forms of thought borrowed from the modern Science of Nature. For the receiver of knowledge thus imparted will then have to apply, to the overcoming of certain difficulties of understanding, faculties of soul which would otherwise remain inactive and tend to the above misuse. The popularising of supersensible knowledge, so frequently desired by overzealous and misguided people, should be avoided. The truly earnest seeker does not call for it; it is but the banale, uncultured craving of persons indolent in thought.

In the ethical and social life as well, humanity has reached a stage of development which makes it impossible to exclude all knowledge of the supersensible from public life and thought. In former epochs the ethical and social instincts contained within them spiritual guiding forces, inherited from primaeval ages of mankind. Such forces tended instinctively to a community life which answered also to the needs of individual soul. But the inner life of man has grown more conscious than in former epochs. The spiritual instincts have thus been forced into the background. The Will, the impulses of men must now be guided consciously, lest they become vagrant and unstable. That is to say, the individual, by his own insight, must be in a position to illumine the life in the physical world of sense by the knowledge of the supersensible, spiritual Being of man.

Conceptions formed in the way of natural-scientific knowledge cannot enter effectively into the conscious guiding forces of the ethical and social life. Destined as it is — within its own domain — to bear the most precious fruits, Natural Science will be led into an absolutely fatal error if it be not perceived that the mode of thought which dominates it is quite unfitted to open out an understanding of, or to give impulses for, the moral and social life of humanity. In the domain of ethical and social life our conception of underlying principles, and the conscious guidance of our action, can only thrive when illumined from the aspect of the Supersensible. Between the rise of a highly evolved Natural Science, and present-day developments in the human life of Will — with all the underlying impulses and instincts — there is indeed a deep, significant connection. The force of knowledge that has gone into our science of Nature, is derived from the former spiritual content of man's impulses and instincts. From the fountain-head of supersensible Realities, the latter must now be supplied with fresh impulsive forces.

We are living in an age when supersensible knowledge can no longer remain the secret possession of a few. No, it must become the common property of all, in whom the meaning of life within this age is stirring as a very condition of their soul's existence. In the unconscious depths of the souls of men this need is already working, far more widespread than many people dream. And it will grow, more and more insistently, to the demand that the science of the Supersensible shall be treated on a like footing with the science of Nature.

Printed in the USA
CPSIA information can be obtained
at www.ICGtesting.com
LVHW090856291023
762472LV00002B/6